Copyright © 2016 by Jack Wilkin

All rights reserved

First Edition (2015) Self-published on Lulu.com

Second Edition (2016) Self-published on Lulu.com

ISBN: 978-1-326-66281-3

"Dinosaurs had big ears, but everyone forgot this because dinosaur ears don't have bones."

-Greg, *Over the Garden Wall.*

Allosaurus: A Biography.

Jack Wilkin.

Contents

Introduction 5
Allosaurus: The Animal. 6
History of *Allosaurus* 17
Meet the family. 22
The Mighty Morrison 30
Morrison Monsters 37
The Jurassic World 53
Birth and Development 63
On the hunt 70
Behaviour 78
Paleopathology 82
Death of the dinosaur 91
Excavation and Big Al 2 96
Conclusion 101
References 102
Image Permissions 122
Systematic Index 131
General Index 135

Introduction

All we have today of the mighty dinosaurs are their fossils. It is sometimes easy to forget that these were once living animals, surviving in a world of unimaginable violence. But, if you know what you're looking for, the signs of life are clear. Nowhere is this clearer than with Big Al, a remarkable *Allosaurus* skeleton from Wyoming.

From his nearly complete skeleton palaeontologists can tell his size, approximate age and even the injuries that contributed to his death. In fact, it isn't a stretch to say that we know more about Big Al than just about any other dinosaur.

The following text aims to decipher the sometimes cryptic nature of dinosaur fossils to unravel the story of this animal's life.

Allosaurus: The Animal.

Allosaurus was a genus of large theropod dinosaur that lived from 150 to 145 million years ago during the Late Oxfordian and Early Kimmeridgian stages of the Late Jurassic[1]. Although there were 7 recognised species of *Allosaurus*[2]., the one we will be focusing on for Big Al is *A. fragilis*[3].

Allosaurus fragilis existed in what is now North America, although others are known from Portugal[4]. *A. fragilis* is known primarily from the Morrison Formation with more than 60 skeletons from Wyoming to New Mexico.

The size estimate given by Gregory. S. Paul was 8.5 metres long and 1.7 tonnes in weight[5]. However, other

---

[1] Paul, 2010
[2] Strauss, 2014a
[3] Bates, 2009
[4] Perez-Moreno, 1999
[5] Paul, 2010

sources give higher estimates of up to 12 metres[6]. These higher estimates maybe due to a closely related genus called *Saurophaganax*. Big Al was only 7.9 metres in length meaning that he was only a sub-adult[7] **(Figure 1.a)**.

Allosaurus was a typical large theropod, having a massive skull on a short S-shaped neck, a long tail and reduced forelimbs which ended in three clawed hands[8] **(Figures; 1.b, 1.c & 1.d).**

The forelimbs were only 35% the length of the hindlimbs[9]. Like other theropods, *Allosaurus* had a furcula (wishbone). These were only recognised in 1996 from the Cleveland-Lloyd Dinosaur Quarry and were originally thought to be gastralia but are now considered part of the growth series of the wishbone[10].

Allosaurus had nine vertebrae in the neck, 14 in the back, and five in the sacrum supporting the hips. The number of tail vertebrae is unknown and varied with

---

[6] Selden & Nudds, 2012
[7] Strauss, 2014a
[8] Donald, 1997
[9] Middleton & Gatesy, 2000
[10] Chure, 1996

individuals; James Madsen estimated about 50[11], while others have considered this number too high and suggested 45 or less[12]. Few dinosaur skeletons are ever complete. But, you get so much overlap between different specimens that you can restore the complete skeleton based on multiply incomplete ones **(Figure 1.e)**.

Page 9;
Figure 1.a: The size range of *Allosaurus* and possible synonym *Epanterias* (largest), compared with a human. You can also see Big Al in the chart.

Page 10;
(Top) Figure 1.b: Mounted *A. fragilis* skeleton cast, San Diego Natural History Museum.

(Middle) Figure 1.c: Skull cast of *A. fragilis*, Oklahoma Museum of Natural History.

(Bottom) Fig 1.d: Hand and claws of *A. fragilis*.

---

[11] Madsen, 1993
[12] Paul, 1988

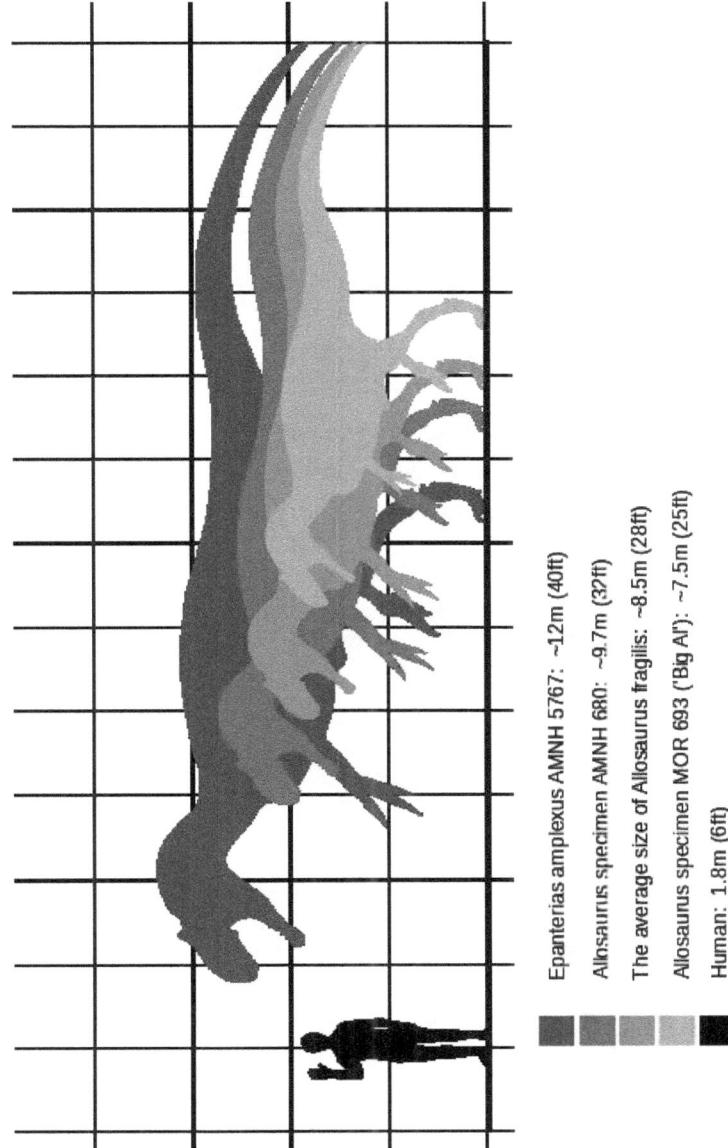

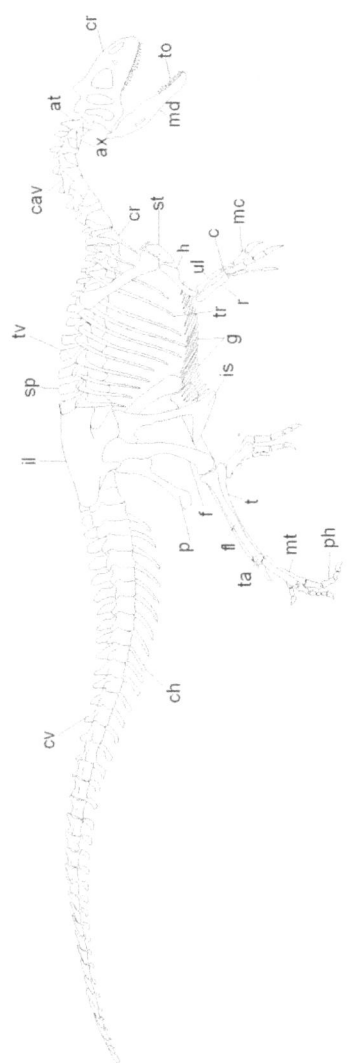

Figure 1.e: Skeletal diagram of *Allosaurus fragilus* adapted from Paul (2010) drawn on CorelDraw7. at= atlas; ax= axis; c= carpal; cav= caudal vertebra; ch= chevron; cr= cervical rib cr= cranium; cv= caudal vertebra; f= femur; fl= fibular; g=gastralia; h=humerus; il= ilium; is- ischium; mc= metacarpals; md=Mandible; mt= metatarsal; p= pubis; ph= phalanges; r=radius; sp= neural spine; st=sternum; t=tibia; ta= tarsals; to= tooth; tr= thoracic rib; tv= thoracic vertebra; ul= ulna.

Authors own work.

Hollow spaces in the neck and anterior back vertebrae have been interpreted as having held air sacs used in respiration[13]. Such air sacs occur in modern birds and are known from theropods. Traditionally air-filled cavities in dinosaur skeletons were viewed just as a way to save weight, but they could have, as Robert Bakker theorised, contained air sacs. The air sacs of birds make the avian respiration system the most efficient of all animals[14].

Allosaurus had a large skull which could measure 85 centimetres in length. The skull is made of two key units; the mandible, (lower jaw) and cranium **(Figure 1.f & g)**. The skull is comprised of many different bones **(Figure 1.h)**. The most important bones to focus on are those that hold the teeth. On the mandible, it's the dentary and on the cranium it's the premaxilla and the maxilla[15]. The skull contained 5 premaxillary teeth, 15-16 maxillary teeth and 17-18 dentary teeth[16]. The teeth of theropods were polyphyodont, meaning they were constantly being

---

[13] Madsen, 1993
[14] Bakker, 1972
[15] Hone, 2016
[16] Currie, 2003

replaced and serrated[17]. The nasals lie behind the maxilla and surround the nares, the opening to the nostrils. Behind these are the antorbital fenestra and then the orbits[18] and behind these the temporal fenestra. Like all theropod, *Allosaurus* skulls contained fenestrae as a way to conserve weight **(Figure 1.i)**.

A pair of horns was situated above and in front of each eye. These horns were extensions of the lacrimal bones. They were probably covered in a keratin sheath and may have had many different functions including; acting as sunshades for the eyes[19] and display. A ridge along the back of the skull was used for muscle attachments.

---

[17] Smith *et al*, 2005
[18] Hone, 2016
[19] Madsen, 1993

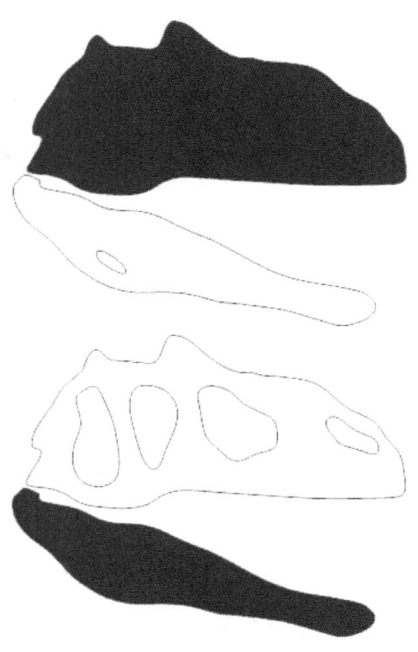

(Top) Figure 1.f: Shaded, upper, part of the skull is the cranium.

(Bottom) Figure 1.g: Shaded, lower, part of the skull is the mandible.

Both diagrams are drawn using CorelDraw7 and are based on the *Allosaurus* skull replica in Burnaby Building at the University of Portsmouth. Authors own work

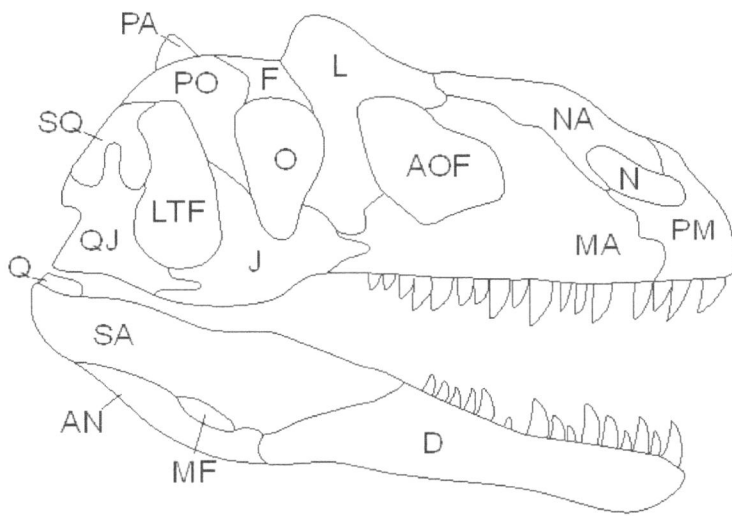

Figure 1.h: Diagram of *Allosaurus* skull based on the specimen in the University of Portsmouth.

Key for labels; AN- Angular; AOF- Antorbital fenestra; D- Dentary; F- Frontal; J- Jugal; L- Lachrymal; LTF- Lower temporal fenestra; MA- Maxilla; MF- Mandibular fenestra; N- Nostrils; NA- Nasal; O- Orbit; PA- Parietal; PM- Premaxilla; PO- Postorbital; Q- Quadrate; QJ- Quadratojugal; SA- Surangular; SQ- Squamosal

Authors own work

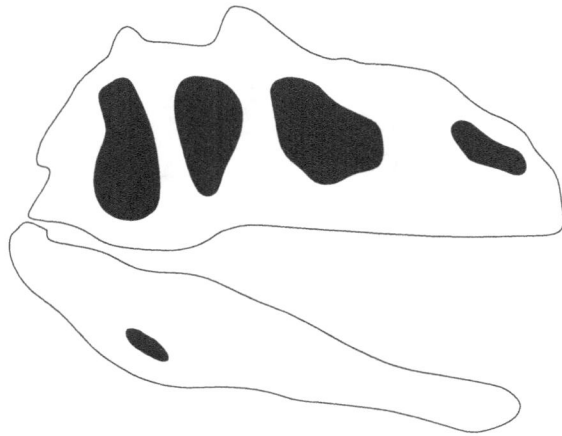

Figure 1.i: Diagram of *Allosaurus* skull showing the locations of holes in the skull. Based on the specimen at the University of Portsmouth. Authors own work

History of *Allosaurus*

The early history of *Allosaurus* is complicated by the multiplicity of names coined during the 'Bone Wars', a thirty-year-long rivalry between American palaeontologists Othniel Charles Marsh and Edward Drinker Cope from 1877-92 **(Figure 2.a)**. It resulted in the discovery of over 130 dinosaur species[20]. However, out of these species, only 32 are recognised as valid today.

Figure 2.a: Marsh (left) and Cope (right).

---

[20] PBS, 2014

Allosaurus was first documented in 1869 when Ferdinand Vandiveer Hayden was presented with a 'petrified horse hoof' by locals from the Middle Park region of Colorado. These fossils were assigned by palaeontologist Joseph Lindy as belonging to the European species *Poekilopleuron* in 1870. Later in 1873, it was assigned its own genus *Antrodemus*[21] which means 'body cavity'.

The discovery of more diagnostic fossils was later unearthed in Colorado and described by Marsh in 1877. He named the new material *Allosaurus fragilis*, meaning fragile different lizard'[22]. The etymology of the name is Greek with *Allo* meaning "different", *sauros* meaning "lizard" and *fragilis* meaning "fragile". Despite this, the name *Allosaurus* only began to be systematically used in the mid-1970's[23]. Since then and now thousands of specimens have been recovered from the Morrison Formation.

---

[21] Breithaupt, 2001
[22] Ibid
[23] Strauss, 2014a

Allosaurus has three synonyms that are no longer used;

- *Creosaurus* (Marsh, 1878).
- *Epanterias*? (Cope, 1878).
- *Labrosaurus* (Marsh, 1879).

Marsh's great contribution to dinosaur palaeontology is that he found dinosaurs on a massive scale. He named many of the largest and most familiar dinosaurs we know today, ones commonly seen in museum halls around the world. For example; Marsh found and described *Allosaurus*, *Apatosaurus*, *Diplodocus*, *Ornithomimus*, *Stegosaurus*, *Triceratops* and much more.

The real work on *Allosaurus* did not begin until the Cleveland-Lloyd excavations of the 1960's. Between 1960 and 1965 thousands of bones were uncovered from the quarry. However, by this point dinosaur research had lost the vigour of the days of Marsh and Cope. There was less fieldwork and fewer dramatic finds. Many dinosaur-bearing formations in Europe had be exhausted and North America, once the treasure trove of the palaeontological world, was producing much of the same material. This isn't to say that fieldwork and innovations did not continue. For example, American Museum of Natural History led expeditions to the Gobi Desert under Roy Chapman Andrew (one of the many inspirations for Indiana Jones) and a lot of work was conducted by Soviet scientists, mostly from Poland, in Mongolia[24].

Luckily hope was around the corner. The Dinosaur Renaissance of the late 1960's reenergized interest in these amazing animals. The scientific and public perception of dinosaurs changed. They went from slow, cold-blooded and deserving of extinction to energetic, somewhat intelligent and warm-blooded. This era also saw new techniques being neutralised and studies were taken to explore their biology, ecology and behaviour. This is different to early palaeontologists, who only focused on the description of the bones and trying to place organisms in the seemly endless tree of life.

As one of the first well-known theropod dinosaurs, *Allosaurus* is known outside of palaeontological circles. *Allosaurus* has appeared in several films and documentaries about prehistoric life. For example, it was

---

[24] Hone, 2016

the top predator in the 1912 novel *The Lost World* by Sir Arthur Conan Doyle. *Allosaurus* appeared in the Westerns *The Beast of Hollow Mountain* (1956) and *The Valley of Gwangi* (1969). In less fictional settings *Allosaurus* has also featured in countless documentaries including (but not limited to); *When Dinosaurs Ruled* (1999), *Walking with Dinosaurs* (1999), and *Dinosaur Revolution* (2011). Big Al himself is not afraid of being in the media spotlight after having an entire *Walking with Dinosaurs* special, *The Ballad of Big Al* (2001), dedicated to him. *Allosaurus* is also the state dinosaur of Utah[25].

---

[25] Brusatte, 2002

Meet the family.

Allosaurus was part of a superfamily of theropods called Allosauroidea. This group contains four families **(Figure 3.a)**;

- Metriacanthosauridae,
- Allosauridae,
- Carcharodontosauridae,
- Neovenatoridae.

The oldest-known allosauroid, *Shidaisaurus* [26], appeared in the early Middle Jurassic of China. The last known definitive surviving members of the group died out around 93 million years ago in Asia, *Shaochilong*[27], and South America, *Mapusaurus.* However, fossil skull fragments have been found dating to the Late Cretaceous Campanian-Maastrichtian Marilia Formation of Brazil but

---

[26] Wu *et al,* 2009
[27] Brusatte *et al*, 2010

these remains are very fragmentary and are yet to be named[28].

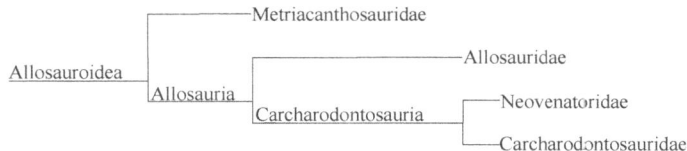

Figure 3.a: Cladogram of Allosauroidea with all four families shown.

Metiacanthosauridae

Metiacanthosaurids were large theropods and the most basal allosauroids. The family is synonymous with Sinraptoridae, but has priority over it[29]. The families name-sake, *Metriacanthosaurus*, is known from the Oxford Clay, Middle Jurassic, of England. The genus name means "moderate spined lizard" as the neural spines were tall for a carnosaur[30].

---

[28] Fernandes de Azevedo *et al*, 2013
[29] Naish & Martill, 2007
[30] Walker, 1964

Yangchuanosaurus from the Late Jurassic of China is the best-known member of the family. It lived at the same time as *Allosaurus* and the roughly the same size, about 8 metres in length[31], though the species *Y. magnus* may have reached 10.8 metres[32]. Phylogenetic analysis conducted by Carrano *et al* found that *Yangchuanosaurus* was a basal member of Metiacathosauridae and a sister group to Metriacanthosaurinae **(Figure 3.b).**

Sinraptor dongi is known the late Middle Jurassic of China[33]. It shared its environment with the 5 metre-long *Monolophosaurus jiangi* [34] and the basal tyrannosaur *Guanlong wucaii* [35]. Despite its etymology, the genus *Sinraptor* is not related to dromaeosaurs.

---

[31] Zhiming, 1975
[32] Paul, 1988
[33] Yi-Ming *et al*, 2013
[34] Zhao & Currie, 1993
[35] Xu *et al*, 2006

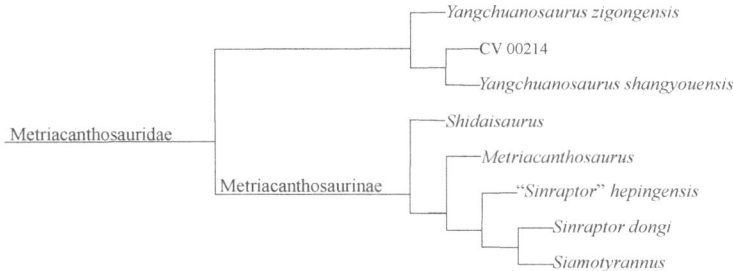

Figure 3.b: Cladogram of Metriacanthosauridea showing all genera.

Allosauridae

Allosauridae was a family of medium to large carnivorous dinosaurs. Members included *Saurophaganax*, *Allosaurus*, and the poorly known *Epanterias*[36]. However, these might be synonymous with *Allosaurus*. Allosauridae was described by American palaeontologist Othniel Charles Marsh in 1878 (Marsh, 1878).

Saurophaganax is a possible allosaurid genus or a species within the genus *Allosaurus*[37]. It was larger than *Allosaurus* with some size estimates giving it a length of

---

[36] Carrano *et al*, 2012
[37] Smith, 1998

13 metres. It is known from Oklahoma and became the official state dinosaur for Oklahoma in 2000 [38] . *Saurophaganax* was large for an allosaurid, and bigger than both *Allosaurus fragilis* and the megalosaur *Torvosaurus tanneri*. Due to its large size, it is likely that *Saurophaganax* was able to hunt giant sauropods.

Whatever the exact number of genera, one thing is clear, the allosaurids were the dominant predators. During the Late Jurassic, they were the most successful theropods on the planet, outnumbering both the ceratosaurs and megalosaurs they competed with for food.

This success was not to last. The allosaurids would eventually be succeeded by their close relatives the carcharodontosaurians (and non-related abelisaurs), in the southern hemisphere, and replaced by the coelurosaurian tyrannosaurs in the northern hemisphere, during the Cretaceous Period.

---

[38] Brusatte, 2002

Neovenatoridae

Diagnostic features of the group include certain features of the vertebrae, a short broad scapula, ilium perforated by numerous cavities and modifications to the femur and tibia[39].

Some phylogenetic studies [40] have concluded that neovenatorids included Maniraptora. However, more recent studies have concluded that Maniraptora were tyrannosauroids [41]. If Maniraptora were part of Neovenatoridae then it would mean that allosauroids lasted until the Cenomanian [42] Stage of the Late Cretaceous. The Maniraptora includes *Orkoraptor*, the most southernmost theropod known from South America[43]. They also include one of the few Japanese dinosaurs, *Fukuiraptor kitadaniensis*. *Fukuraptor* was smaller than most other allosauroids at 4.2 metres long[44].

---

[39] Benton, 2014
[40] Benson *et al*, 2010
[41] Novas *et al*, 2012
[42] Varela *et al*, 2012
[43] Novas *et al*, 2008
[44] Currie & Azuma, 2006

Neovenator, was discovered on the Isle of Wight in 1996. It was the apex predator of Northern Europe during the Early Cretaceous. It could reach 7.5 metres in length and 2.5 metres in height[45].

Carcharodontosauridae

The name carcharodontosaurid means "shark toothed lizards" because their teeth somewhat mirror those of sharks. They appeared later than the Allosauroidae and are the most derived group of allosauroid **(Figure 3.c)**. They ranged from the Late Jurassic[46] to the Campanian. During the latter part of the Cretaceous, they were restricted to South America [47]. Carcharodontosaurs were mainly Gondwana with species from Brazil[48], Argentina[49] and Africa[50,51]. Despite this overwhelming preference to the

---

[45] *Neovenator salerii,* 2016
[46] Rauhut, 2011
[47] Fernandes de Azevedo *et al*, 2013
[48] Fernandes de Azevedo *et al*, 2013
[49] Coria & Salgado, 1995
[50] Rauhut, 2011
[51] Brusatte & Sereno, 2007

southern continents, carcharodontosaurs are also known from North America with the Early Cretaceous *Acrocanthosaurus*[52] and *Shaochilong* of China[53].

Carcharodontosaurids include some of the largest theropods to ever live. The genera *Giganotosaurus* of Argentina and *Carcharodontosaurus* of Africa could reach approximately 13 metres in length [54]. These estimates rival those of *Tyrannosaurus rex* whose largest specimen, Sue, is 12.3 metres in length[55].

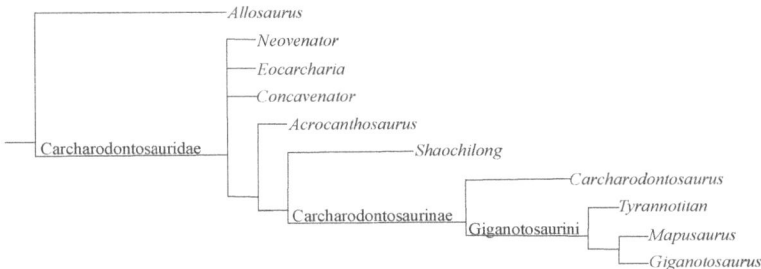

Figure 3.c: Cladogram of Carcharodontosauridae showing all genera with *Allosaurus* as an outlier.

---

[52] Eddy & Clarke, 2011
[53] Brusatte *et al*, 2010
[54] Therrien & Henderson, 2007
[55] Hone, 2016

The Mighty Morrison

Big Al was found in the Upper Jurassic Morrison Formation (Bates, 2009) like most other *Allosaurus* fossils. The formation covers 600,000 square miles and runs from Alberta to New Mexico and from Idaho across to Nebraska[56] **(Figure 4.a)**.

The Morrison dates to the Late Jurassic 155-148 million years ago[57]. It was dated using radiometric dating and biostratigraphy[58]. The Morrison Formation is also a major source of uranium ore. In fact, the radiometric dating done on the Morrison Formation used uranium-led (U-Pb) dating and dated it to the Oxfordian Stage[59].

---

[56] Haines, 1999
[57] National Park Service, 2014
[58] Selden & Nudds, 2012
[59] Trujillo, 2006

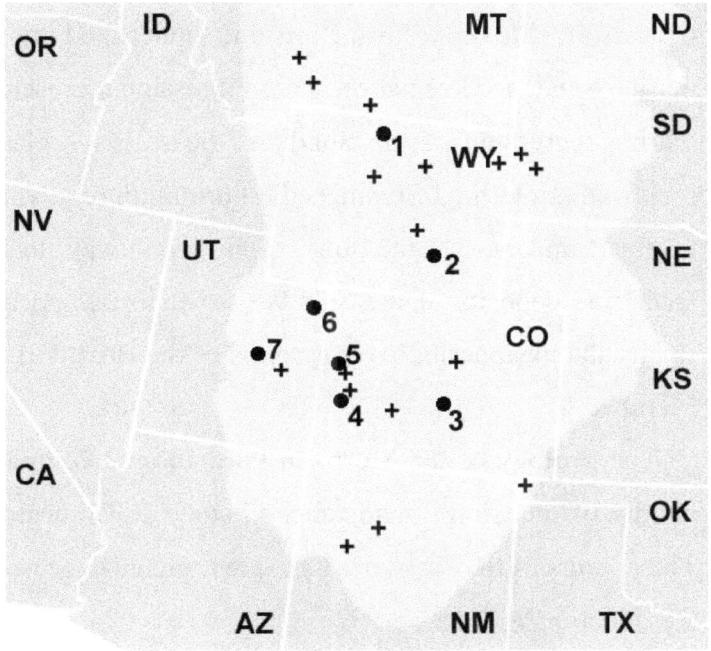

1) "Big Al" quarry, Big Horn Co., WY.
2) Como Bluff, Albany Co., WY.
3) Garden Park/Cañon City, Fremont Co., CO.
4) Dry Mesa Quarry, Delta Co., CO.
5) Grand Junction/Fruita, Mesa Co., CO.
6) Dinosaur National Monument West, Uintah Co., UT.
7) Cleveland-Lloyd Dinosaur Quarry, Emery Co., UT.

Figure 4.a: A map of the Morrison Formation which also shows the locations of *Allosaurus* quarries including Al.

The majority of the Morrison is capped by the Lower Cretaceous Cloverly Formation and underlined by the marine Sundance Formation. This succession represents a marine regression of the Sundance Sea. This is clearly seen in the intertidal Summerville Formation (equivalent to the Sundance Formation) as it gives way to the lacustrine deposits of the Salt Wash Member, and then finally the meandering river systems of the Brushy Basin Member[60].

The geology of the Morrison Formation is composed mainly of mudstone, sandstone, siltstone and limestone. The colours of the rocks are light grey, greenish grey and red due to paleosols[61].

The Morrison Formation includes a range of environments from wet swamps complete with coal deposits in the north to desert conditions in the south[62].

Most of the fossils come from the green siltstone beds and lower sandstones. The States of Colorado, Utah and Wyoming contain mostly fluviatile and lacustrine

---

[60] Selden & Nudds, 2012
[61] Geoweb, 2015
[62] Selden & Nudds, 2012

deposits[63]. Many of the fossils are partly dislocated due to transportation by rivers before burial in sandbars. Freshwater species suggests that there were fresh water lakes as well as saline, alkaline lakes[64].

The Morrison is notable for its dinosaur fossils which include *Diplodocus, Apatosaurus, Brachiosaurus, Dryosaurus* and *Stegosaurus* just to name a few herbivores. Theropods at the formation include *Torvosaurus, Coelurus* and *Ceratosaurus*[65]. It is in the Midwestern States of Colorado, Utah and Wyoming are richest deposits can be found. Here flash floods have deposited thousands of tonns of bones in what is known as a Concentration Lagerstätte.

The Morrison Formation is probably best known for its dinosaur bone beds such as Dinosaur National Monument and the Cleveland-Lloyd Dinosaur Quarry, both in Utah. A bone bed is a sedimentary deposit that contains remains of multiple individuals in a concentration greater than the

---

[63] Selden & Nudds, 2012
[64] Tang, 2014
[65] Sampson, 2009

background level[66]. Dinosaur National Monument is a 329.44 square mile national park in Utah which was established in 1915[67]. The park's geological history spans 1.1 billion years, but it's the Late Jurassic sediments we are interested in. The Morrison Formation in the park represents stream, lake and swamp deposits. The quarry in the park has a range of dinosaur species including *Diplodocus, Apatosaurus, Stegosaurus* and *Allosaurus*[68]. The quarry has a tilted rock layer that contains hundreds of dinosaur fossils[69] **(Figure 4.b)**. The quarry is now enclosed by the Dinosaur Quarry Building to protect it. The rock layer enclosing the fossils is a sandstone and conglomerate bed of alluvial or river bed origin. The dinosaurs and other ancient animals were washed into the area and buried presumably during flooding events **(Figure 4.c)**.

The Cleveland-Lloyd Dinosaur Quarry in Utah has some of the best evidence of theropod dinosaurs anywhere

---

[66] Rogers *et al*, 2007
[67] National Park Service, 2015
[68] Foos & Hannibal, 1999
[69] Sampson, 2009

in the world. It is the largest assemblage of theropod dinosaurs in the world as it has yielded over 50 *Allosaurus* individuals representing different growth stages. *Allosaurus* is the predominant species at the site, although other theropods are known but in far fewer numbers. Herbivorous dinosaurs are also present but are very rare[70]. This site has been interpreted as a predator trap due to the 3:1 ratio of predators to prey. The most realistic, given the data, fortheory for the sites formation is the drought hypothesis. Evidence includes a large assemblage of animals in a low energy ephemeral-pond depositional setting and geological and biological evidence of desiccation[71].

The Morrison gives a remarkable insight into Late Jurassic ecosystems that includes not only some of the largest dinosaurs of all time but also other animals that coexisted alongside them[72]. Specimens from the Morrison Formation can be seen in museums all over the world such as the Natural History Museum in London, the American

---

[70] Sampson, 2009
[71] Gates, 2005
[72] Selden & Nudds, 2012

Museum of Natural History in New York and the Royal Ontario Museum in Toronto.

Figure 4.b: Workers working in the Dinosaur Quarry.

Figure 4.c: The distinctive banding of the Morrison Formation originated from muds and sands laid down by ancient rivers.

Morrison Monsters

The Morrison Formation is renowned worldwide for being one of the greatest dinosaur fossil formations in the world. The following chapter is a quick tour of the different dinosaur groups that are known from the Morrison.

Part 1: Ornithischians

The herbivorous ornithischian dinosaurs are diverse, but not as common as sauropods, in the Morrison. This group of dinosaurs are often called 'bird-hipped' because the arrangement of the hip bones is like that of birds with the pubis pointing backwards[73] **(Figure 5.a)**. Ironically birds evolved from the lizard-hipped saurischians and not the bird-hipped dinosaurs.

---

[73] Paul, 2010

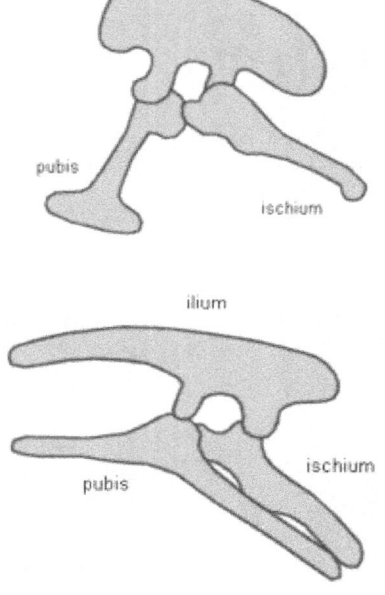

Figure 5.a: The different arrangements of dinosaur hip bones. Top is the saurischian (theropods and sauropods) with the pubis pointing forward like in reptiles. The diagram on the bottom is the ornithischian hip with the pubis pointing backwards.

1.1: Ornithopods

Ornithopods were bipedal herbivores that came in several types. Small "hypsilophodonts" included *Drinker*, *Nanosaurus* and *Othnielia*. Larger but similar-looking dryosaurids were represented by *Dryosaurus* and the camptosaurid by *Camptosaurus* **(Figure 5.b)**. Both Dryosaurids and camptosaurids were early iguanodonts, a group that would later spawn the duck-billed dinosaurs. Duck-billed dinosaurs (or hadrosaurs) were the most successful group of dinosaurs in the Cretaceous. As well as bones, ornithopod eggs shells have also been found in the Morrison[74].

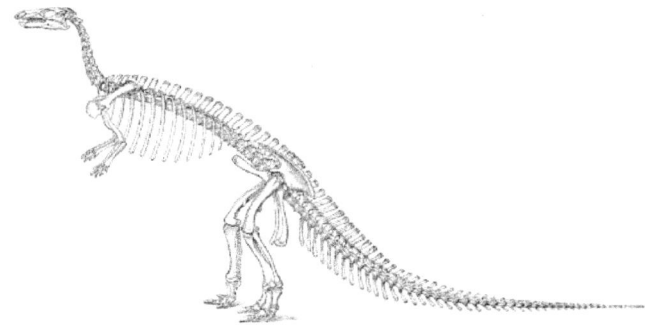

Figure 5.b: Historical skeletal restoration of *Camptosaurus* by O.C. Marsh.

1.2: Thyreophorans

This group of dinosaurs are often called 'armoured dinosaurs' due to their armour plating. The most famous of these from the Morrison is *Stegosaurus*. It had a row of plates running down its back that may have been used as

---

[74] Selden & Nudds, 2012

a temperature-regulating device[75]. Its tail sported 4 spikes called thagomizers which would have been used for defence **(Figure 5.c)**. At least three species have been identified in the upper Morrison Formation and are known from the remains of about 80 individuals[76].

Armoured dinosaurs that weren't stegosaurs were unknown in the formation until the 1990s[77]. Two have been named: *Gargoyleosaurus* and *Mymoorapelta*. Both of these are ankylosaurs with *Gargoyleosaurus* being one of the earliest ankylosaurs represented by reasonably complete fossils.

Figure 5.c: Life reconstruction of *Stegosaurus*.

---

[75] Selden & Nudds, 2012
[76] Turner & Peterson, 1999
[77] Foster, 2007

1.3: Heterodontosauridae

Heterodontosaurs are primitive ornithischian dinosaurs which first evolved in the Late Triassic (around 215 mya), and continued until the Early Cretaceous (around 133 mya). Heterodontosaurids are named for their strongly heterodont dentition. Many species had a large third tooth that resembles the canines of carnivoran mammals and are often called tusks.

The only species of heterodontosaur known from the Morrison is *Fruitadens*. It was a small species and is the smallest dinosaur from the Morrison at 75 cm in length **(Figure 5.d)**. It may, as well with other heterodontosaurids, have been omnivorous. A 2012 study suggested *Fruitadens* was an ecological generalist, eating both plants and possibly insects or other invertebrates[78].

---

[78] Butler, 2012

Figure 5.d: Life reconstruction of Fruitadens.

Part 2: Sauropods

The name Sauropoda was coined by O.C. Marsh in 1878, and is derived from Greek, meaning 'lizard foot'[79]. They were the largest dinosaurs, and land animals, ever to have existed. Complete sauropod fossil finds are rare. Many species, especially the largest, are known only from isolated and disarticulated bones.

---

[79] Marsh, 1878

Many near-complete specimens often lack heads, tail tips and limbs. Sauropods were herbivorous, long-necked and had pillar-like legs. They had tiny heads, massive bodies and most had long tails[80]. They would have had a massive impact on the environment.

Some species could have used their tails like a whip. Some palaeontologists have suggested that diplodocid tails could have been used as defensive weapons. More recently, several have speculated that, like bullwhips, such tails were noisemakers used for communication[81].

2.1: Diplodicoids

The gigantic diplodocids are known from the upper Morrison. They include the genera *Diplodocus* (formerly *Seismosaurus*), *Supersaurus*, and the largest of all, *Amphicoelias fragilimus*. Smaller sauropods, such as *Suuwassea* from Montana, tend to be found in the northern reaches of the Morrison, near the shores of the

---

[80] Tidwell *et al*, 2001
[81] Peterson, 2002

ancient Sundance Sea. This suggests ecological niches favouring smaller body size in the north as compared with conditions favouring gigantism further south [82]. *Amphicoelias fragillimus* could well be the largest dinosaur ever and the longest ever vertebrate at 60 metres long **(Figure 5.e)**. However, the fossils of *A. fragillimus* were lost in the 1870's.

Diplodocus **(Figure 5.f)** is one of the most common dinosaurs from the Morrison and one of the best-known sauropods. *Diplodocus* was a large dinosaur at 27 metres in length[83]. The neck was composed of 15 vertebrae and was positioned more or less horizontal to the ground[84].

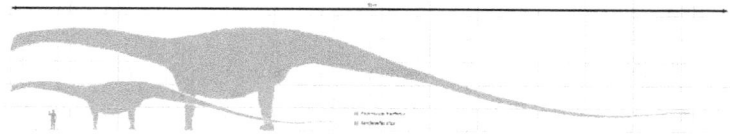

Figure 5.e: *A. fragillimus* **(big dinosaur) and** *A. altus* **(smaller dinosaur) compared in size with a human.**

---

[82] Harris & Dodson, 2004
[83] Selden & Nudds, 2012
[84] Stevens, 1999

Figure 5.f: Life reconstruction of *Diplodocus*.

2.2: Apatosaurines

This is a subfamily within Diplodocidae. It includes the genus *Apatosaurus,* formally known as *Brontosaurus*. They are similar to diplodocids but have a bulkier skeleton. *Diplodocus* only weighed about 12 tonnes, but *Apatosaurus* would have been 20 tonnes even though it was 7 metres shorter[85] **(Figure 5.g)**.

---

[85] Selden & Nudds, 2012

Figure 5.g: Life reconstruction of *Apatosaurus* arching its neck down to drink.

2.3: Macronarians

The middle stages of the Morrison Formation were dominated by familiar forms such as the Giraffe-like *Brachiosaurus*, and related camarasaurids, like *Camarasaurus*. Even though *Brachiosaurus* could not hold its neck up vertically, the head height would still have been 9.4 metres off the ground[86] **(Figure 5.h)**. This giant dinosaur fed primarily on ginkgos, conifers, tree

---

[86] Foster, 2007

ferns, and large cycads and would have required around 240 kilogrammes per day[87]. *Camarasaurus* was more massive than *Diplodocus* at 18 tonnes. *Camarasaurus* is nicknamed the 'Jurassic cow' due to their abundance[88].

Figure 5.h: Life reconstruction of *Brachiosaurus*.

Part 3: Theropods

Theropods, like the previously discussed sauropods, are saurischian dinosaurs. The group Theropoda was primarily carnivorous, although a number of theropod

---

[87] Hummel, 2008
[88] Selden & Nudds, 2012

groups evolved herbivory, omnivory, and insectivory diets. Theropods were the first dinosaurs as they appeared during the Carnian age of the Late Triassic period about 230 million years ago. Theropods were the sole large terrestrial carnivores from the Early Jurassic until at least the close of the Cretaceous. They are the only group of dinosaurs that are alive today, as birds evolved from small specialised coelurosaurian theropods in the Jurassic. Today they are represented by between 9000 to 10,000 living species[89].

3.1: Carnosaurs

All the carnosaurs at the Morrison Formation are represented by the family Allosauridae, which has already been discussed in the chapter "Meet the Family". *Allosaurus* was the most common large theropod in the Morrison Formation, accounting for 70 to 75% of theropod specimens[90].

---

[89] Klappenback, 2014
[90] Foster, 2007

3.2: Ceratosaurs

Ceratosaurs were less derived than allosaurids. This group is defined as all theropods that share a more recent common ancestry with *Ceratosaurus* than with birds. There is no agreed listing of species or diagnostic characters of this group. Ceratosauria appeared in the Late Jurassic and went extinct Late Cretaceous. In the Cretaceous, this group was found primarily (though not exclusively) in the Southern Hemisphere.

Two genera are known from the Morrison; *Ceratosaurus* and *Elaphrosaurus*. *Ceratosaurus* could reach 6 metres in length. It had a nasal horn that may have been for display and if so it would have been brightly coloured [91]. Interestingly, *Certatosaurus* had a long, flexible body, with a deep tail shaped like that of a crocodilian. This means that this dinosaur could have, as suggested by Robert Bakker, hunted aquatic prey such as fish and crocodiles [92]. *Elaphrosaurus* could reach 6.2

---

[91] Foster, 2007
[92] Bakker & Bir, 2004

metres long and was more lightly built than *Certatosaurus*.

3.3: Coelurosaurs

Coelurosaurs are highly derived theropods closely related to birds. The clade Coelurosauria is a subgroup of theropod dinosaurs that includes compsognathids, tyrannosaurs, ornithomimosaurs, and maniraptorans. The clade Maniraptora includes birds, the only dinosaur group alive today [93]. Many specimens of coelurosaurs are feathered and it is probable that the entire group was feathered[94].

It was during Late Jurassic that the tyrannosaurs first evolved. This group remained in the shadows of giant allosaurids throughout the Late Jurassic into the early part of the Cretaceous. However, they would evolve into giant carnivores like *Tarbosaurus* and *Tyrannosaurus*. Tyrannosauroidea is represented in the Morrison by *Coelurus, Stokesosaurus* and *Tanycolagreus* **(Figure 5.i)**.

---

[93] Turner *et al*, 2012
[94] Currie, 2005

Figure 5.i: Life reconstruction of *Tanycolagreus*, a possible early tyrannosaur.

3.4: Megalosaurids

They were relatively primitive tetanuran theropods. Megalosaurids first appeared in the Middle Jurassic and seemed to have been displaced and replaced by other theropods (allosaurs) by the end of that period. The Middle Jurassic megalosaur genus *Megalosaurus* was the first dinosaur to be scientifically described. It was described in 1824 by Sir William Buckland. The dinosaur came from Oxfordshire, Southern England.

The genus *Torvosaurus* is known from the Morrison through a series of incomplete specimens[95] so was likely much rarer than *Allosaurus*. It was a large predator comparable to *Allosaurus* in size at 9 metres in length[96]. The species *T. gurneyi* discovered in Portugal is the largest theropod known from Europe at 11 metres long[97].

---

[95] Sampson, 2009
[96] Paul, 2010
[97] Hendrickx, 2014

The Jurassic World

The Jurassic marked a transition in the evolution of the planet. The Triassic-Jurassic extinction left many ecological niches empty. This led to dinosaurs becoming dominate in every niche on land [98]. This extinction occurred roughly 202 million years ago and was one of the 'big five' mass extinctions of the Phanerozoic. The extinction occurred during a hothouse with extremely high CO_2 levels and the supercontinent of Pangea had not yet broken up leading to large arid regions in the interior. In continental areas, the extinction marked the end of a period dominated by non-dinosaurian tetrapods and the start of the Age of the Dinosaurs[99].

The Early Jurassic was still mostly dry, but as the era progressed the climate became wetter and wetter [100].

---

[98] Haines, 1999
[99] Koeberl & MacLeod, 2002
[100] Haines, 1999

During this time coal deposits formed in Australia and even Antarctica[101].

In the Late Jurassic the continents were arranged into two landmasses after the breakup of Pangea; Laurasia in the north and Gondwana in the south [102] **(Figure 6.a)**. *Allosaurus* lived in Laurasia. Sea levels rose, creating epicontinental seaways in North America and Europe[103]. Interestingly *Allosaurus fragilis* is the first dinosaur species to be recognised on two different continents (North America and Europe). This suggests that a land bridge may have existed between the two landmasses during the Late Jurassic[104].

---

[101] Cantrill, 2012
[102] Ezcurra & Agnolín, 2012
[103] Smithsonian, 2014
[104] Perez-Moreno, 1999

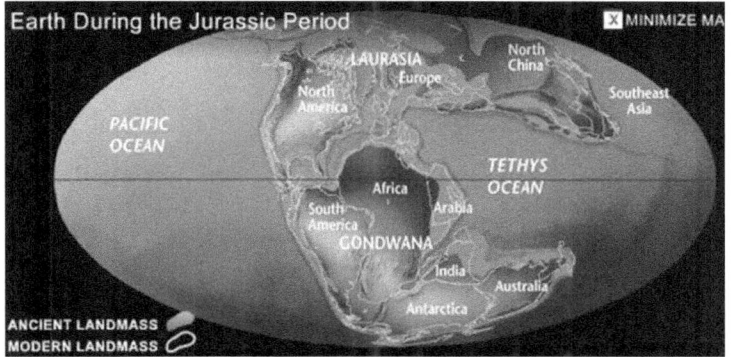

Figure 6.a: Map of earth during the Late Jurassic.

The dominant plants during the Jurassic were ferns, conifers, cycads and ginkgoes. Insects became more diverse in the Jurassic with some groups involved in plant reproduction by transferring pollen from one plant to another. It is generally accepted that flowering plants (Angiosperms) evolved in the Early Cretaceous [105]. However, there is some evidence from China that suggests that they appeared in the Late Jurassic[106] though this date has since been revised to the Lower Cretaceous[107].

---

[105] Taylor *et al*, 2009
[106] Sun *et al*, 1998
[107] Ji *et al*, 2004

The breakup of Pangea had a profound effect on marine life by bringing previously isolated communities together. The increased amount of continental shelf submerged caused by sea-level rise and a greenhouse climate provided more iterations and chances to evolve, resulting in increasing diversity in life in the oceans. Also, the evolution of shell-crushing behaviour among Mesozoic marine predators forced shelled organisms to develop defences against such predation. This increase in diversity is known as the Mesozoic Marine Revolution[108].

Ammonites were a type of cephalopod related to today's squids and nautiloids. They first appeared before the dinosaurs in the Devonian but went extinct along with them at the end of the Cretaceous. Due to their wide distribution and rapid evolution they are often used as zone fossils in biostratigraphy. The majority of ammonite species have coiled shells, but in some species, the shell has uncoiled. These uncoiled types are called

---

[108] Vermeij, 1977

heteromorphs **(Figure 6.b)**. Closely related to ammonites are the bullet shaped belemnites[109].

Figure 6.b: A reconstruction of the heteromorphic ammonite *Hamites*.

Fish and sharks were common in the Jurassic seas as they are today. The pachycormid fish *Leedsichthys* was the biggest fish ever. Early estimates put the fish's length at 27.6 metres for the largest individuals[110] though more recent studies have cut that down to 16.7 metres[111]. In life

---

[109] Clarkson, 1993
[110] Martill, 1988
[111] Switek, 2013

this fish acted like a baleen whale, filtering out plankton using its gill racks[112].

Marine reptiles included ichthyosaurs, who were at the peak of their diversity, plesiosaurs, pliosaurs, and marine crocodiles of the families Teleosauridae and Metriorhynchidae[113].

Plesiosaurus had long, flexible necks that they used to catch fish. Pliosaurs had shorter necks and may have fed on fish, sharks, ichthyosaurs, dinosaurs and plesiosaurs. One of the best-known pliosaurs from the Jurassic was *Liopleurodon. Liopleurodon* is found across Europe in what was once the Tethys Sea. The largest species, *L. ferox*, is estimated to have reached a length of 6.5 metres[114].

Ichthyosaurs were marine reptiles that looked very much like dolphins and were the most perfectly adapted for marine life, even going so far as giving birth to live young[115] **(Figure 6.c)**.

---

[112] Benton, 2014
[113] Motani, 2000
[114] Noe *et al*, 2003
[115] Spinar, 1995

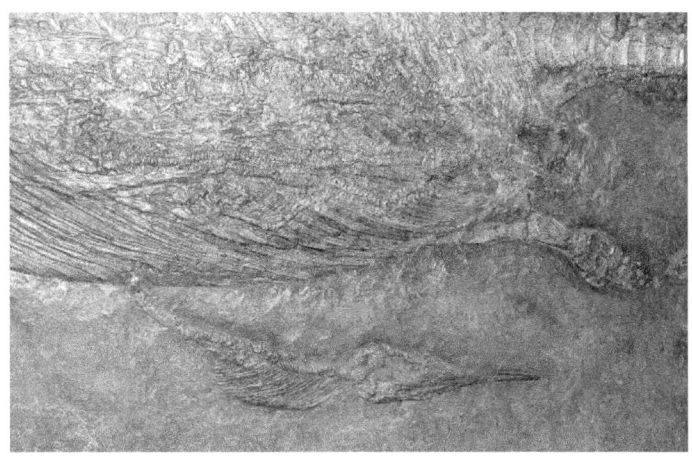

Figure 6.c: This ichthyosaur (*Stenopterygius*) fossil from Germany clearly shows a mother ichthyosaur giving birth to live young.

In the skies above the heads of the dinosaurs were giant flying reptiles called pterosaurs. Like the dinosaurs, they evolved in the Triassic and died out at the end of the Cretaceous 66 million years ago. Despite their size they were able to fly due to hollow bones and their wings were made of a thin skin membrane supported by an extra-long fourth finger[116] **(Figure 6.d)**. Pterosaurs are rare in the

---

[116] Witton, 2013

Morrison Formation due primarily to the fragility of their bones. Pterosaur finds from the Morrison include both rhamphorhynchoids and pterodactyloids[117].

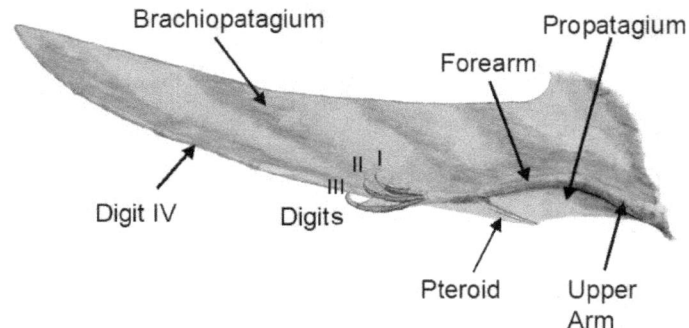

Figure 6.d: The wing anatomy of a pterosaur. Note the exceptionally long fourth digit.

Within the undergrowth below there were various types of mammals. The groups of mammals in the Morrison include; Symmetrodonts, Eutriconodonts, Dryolestoids and Multituberculates. Mesozoic mammals were small and mainly nocturnal. A study in 2009 on these mammals concluded that the genus *Docodon* was the

---

[117] Foster, 2007

largest at 141g and *Fruitafossor* was the smallest at just 6g[118]. Mammals evolved from non-mammalian synapsids (sometimes referred to as mammal-like reptiles), the first true mammals appeared in the Triassic. It was in the Cenozoic, after the extinction of the non-avian dinosaurs, that mammals radiated to become the dominant land animals.

It was also during this time that birds first evolved. *Archaeopteryx* from Solnhofen Limestone of Germany is considered by many to be the first bird. It had avian features such as wings and feathers but also reptilian features like a long bony tail, claws on its hands and teeth[119] **(Figure 6.e)**.

Birds evolved from a group of theropod dinosaurs called Maniraptora[120]. Many features link birds and dinosaurs. Feathers are known from many theropod dinosaur groups including tyrannosaurids[121]. Other features linking the two groups are; hollow bones, air sacs[122], furcula, a

---

[118] Foster, 2009
[119] Benton & Harper, 2009
[120] Chiappe, 2009
[121] Xu *et al*, 2004
[122] Sereno *et al*, 2008

similar egg shell structure [123, 124], and brooding behaviour[125]. Molecular evidence from a *Tyrannosaurus* from Montana shows that birds and dinosaurs are more closely related than dinosaurs and crocodilians[126].

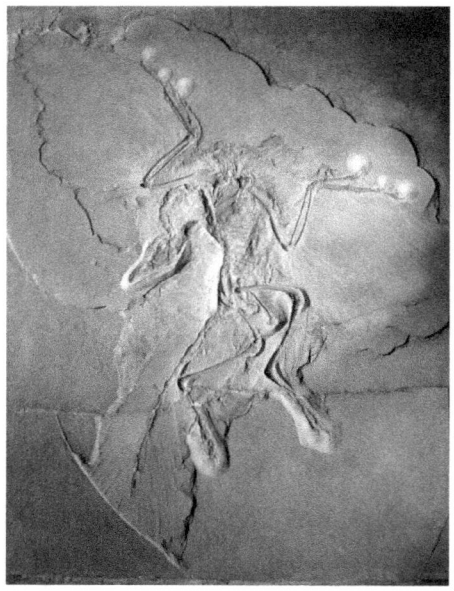

Figure 6.e: A paleontological icon. This is the Berlin specimen of *Archaeopteryx lithographica* from Germany. Note the different avian and reptilian features present in the fossil.

---

[123] Chiappe, 2009
[124] Norell *et al*, 1994
[125] Ibid
[126] Asara *et al*, 2007

Birth and Development

Al, like all dinosaurs, would have come from an egg. The eggs of theropods are generally elongated, mostly between 2 and 4 times longer than they are wide. They were made out of calcite and all contain respiration pores to allow the embryo to breathe[127]. Even Medullary bone tissue has been reported in at least one *Allosaurus* specimen, a shin bone from the Cleveland-Lloyd Dinosaur Quarry[128]. This is evidence that the allosaur in question was female and had a close evolutionary history to birds. A similar discovery was made on the *Tyrannosaurus* specimen 'MOR 1125'[129].

At a site known as Fox Mesa from Wyoming, which is part of the Morrison Formation, a dinosaur nest site was discovered that contained both eggshells and embryonic

---

[127] Larson, 1998
[128] Lee & Werning, 2008
[129] Schweitzer, 2005

bones belonging to *Allosaurus*. The nest was heavily weathered but the egg-shell fragments preserved suggest that the eggs were about 8 x 6.5 cm. The oogenus was *Prismatoolithus coloradensis* and the embryonic remains confirm a theropod origin for the eggs[130].

However, better clues of what an embryonic allosaur might have been like, have been found on the Atlantic coast of Portugal. Under the microscope, palaeontologists have been able to identify holes in the shells. The holes were large to let the maximum amount of air enter, allowing the embryo to breathe. This suggests that the eggs were buried, packed together, in the underground nest. This can be seen today in modern crocodiles. As well as the eggs shells the scientists also discovered fossilised embryos still inside their eggs. They found over 200 embryonic bones[131]. The egg shells in question probably belonged to the megalosaurid *Torvosaurus* and come from the Lourinhã Formation[132].

---

[130] Carrano *et al*, 2013
[131] Haines, 2001
[132] Araujo, 2013

The dinosaur embryos also had teeth[133] **(Figure 7.a)**. They were too small for ripping chunks off the mother's dinner plate, but would have been perfect for catching insects. This meant from a young age Al would have been able to fend for himself. Though the mother would have provided protection. Again, baby crocodiles do the same thing.

Figure 7.a: Illustration of the embryonic *Torvosaurus* snout clearly showing the teeth. Image courtesy of Ricardo Araújo.

---

[133] Araujo, 2013

Allosaurus could live to be 50 or 60 years old[134] and reach over 8 metres long. To reach that size Al would have to grow up fast. They would have been able to grow quickly due to their metabolism.

For most of the early twentieth-century, dinosaurs were seen as slow and sluggish reptiles. But all this changed during the Dinosaur Renaissance of the 1960's which saw a renewal in academic and public interest in dinosaurs. This small-scale scientific revolution was pioneered by John Ostrom and Robert Bakker. Views of dinosaur physiology changed as they began to be viewed as active, bird-like and warm-blooded[135]. The discovery of feathered dinosaurs from China in the mid-1990's helped to affirm this[136,137].

One theory is *that Allosaurus* was not warm-blooded (endothermic) like mammals or birds nor was it cold-blooded (ectothermic) like lizards but may have been mesothermic. This would have given Al, as well as other

---

[134] Young, 2011
[135] Bakker, 1975
[136] Ji & Ji, 1996
[137] Ji & Ji, 1997

dinosaurs, a metabolic rate similar to tuna and great white sharks[138].

When dinosaurs first evolved they competed with cold-blooded reptiles, but because they had a higher metabolic rate than them it meant they could grow faster and become more dangerous predators. However, complete warm-bloodedness would have put Al at a disadvantage as it limits the size an animal can reach. For example, an allosaur sized lion wouldn't be able to hunt enough prey to survive. However, because Al had a lower metabolic rate than lions it meant he had lower food demands. This meant dinosaurs could still get big while maintaining their advantage over the competition[139].

To get the age at which a dinosaur died and then work out its growth rate, palaeontologists make cross sections of the bone. They then count the growth rings, in the bones and compared them to crocodiles and birds. Like trees, bone tends to grow at uneven rates with more bone being deposited during good times and then slowing down when

---

[138] Sullivan, 2014
[139] Sullivan, 2014

time got tough. In thin-section, slowed growth is shown as dark concentric rings called lines of arrested growth, or LAGs for short. The slower bone formation is often liked to seasonal changes[140]. The results show that *Allosaurus* grew like birds, so Al would have grown up very quickly **(Figure 7.b)**. Al would have reached his full size in just 6 to 8 years[141]. Al himself was probably 7 years old[142].

Dinosaurs likely continued growing throughout their lives as do modern crocodilians[143]. This is reflected in the bone microstructure of both dinosaurs and crocodilians. When the animal is young the distance between LAGs was large indicating fast growth but as they got older the distances shortened.

However, *Allosaurus* and some other dinosaur genera have an outer circumferential layer[144]. This forms slowly on the outermost surface of a bone once an animal has

---

[140] Chinsamy-Turan, 2005
[141] Haines, 2001
[142] Breithaupt, 2001
[143] Ricalès, 1980
[144] Reid, 1996

reached its mature body size. It only appears if it has a determined growth strategy[145].

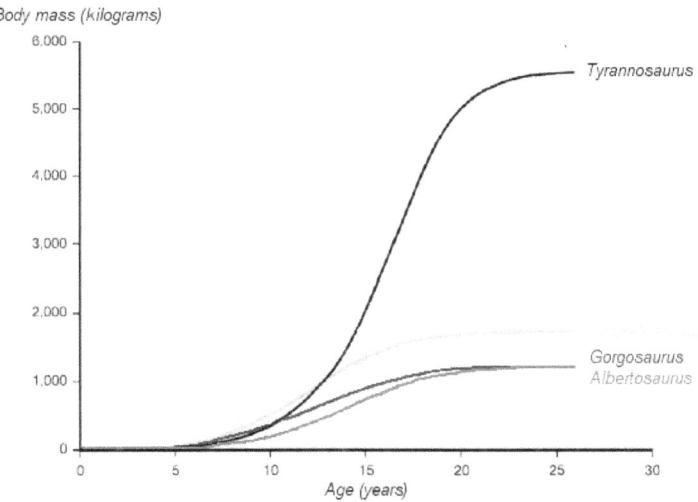

Figure 7.b: This isn't a growth curve for *Allosaurus*, but it still shows how rapidly theropods could grow. The theropods used here are all tyrannosaurids (based on Erickson *et al*, 2004).

---

[145] Chinsamy-Turan, 2005

On the hunt

Allosaurus was a large theropod dinosaur and so Al would have eaten meat...and a lot of it. As previously mentioned when he was a chick Al would have eaten insects and then would have likely progressed on to small vertebrates. However, given how quickly he would have grown up, it wouldn't be long until he was hunting dinosaurs.

There is plenty of evidence to suggest *Allosaurus* hunted the heavily armoured *Stegosaurus*. Such evidence includes a *Stegosaurus* neck plate with a U-shaped wound that correlates well with an *Allosaurus* snout, and an *Allosaurus* tail vertebra with a partially healed puncture wound that fits a *Stegosaurus* tail spike[146] **(Figure 8a & b)**. To add further credence to the theory that stegosaurs used their tail spikes for defensive purposes is that 9.8% of *Stegosaurus* specimens in a 2001 study exhibited injuries on their thagomizers[147].

---

[146] Carpenter *et al*, 2005
[147] McWhinney *et al*, 2001

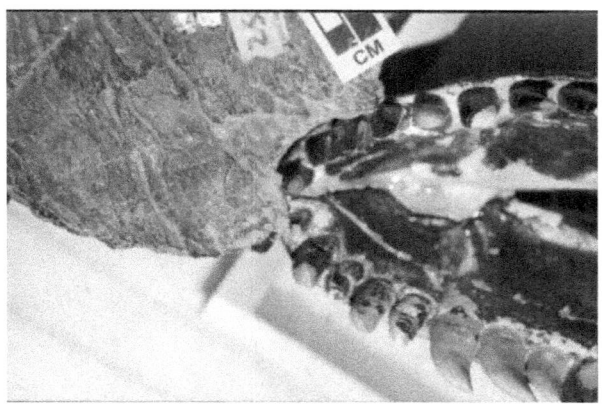

Figure 8.a: *Stegosaurus* **plate showing pathology caused by an** *Allosaurus,* **with an allosaur lower jaw to show the match. Image courtesy of Kenneth Carpenter.**

(Next Page) Figure 8.b: *Allosaurus* **vertebra with a puncture that perfectly matches a stegosaur thagomizer. Image courtesy of Kenneth Carpenter.**

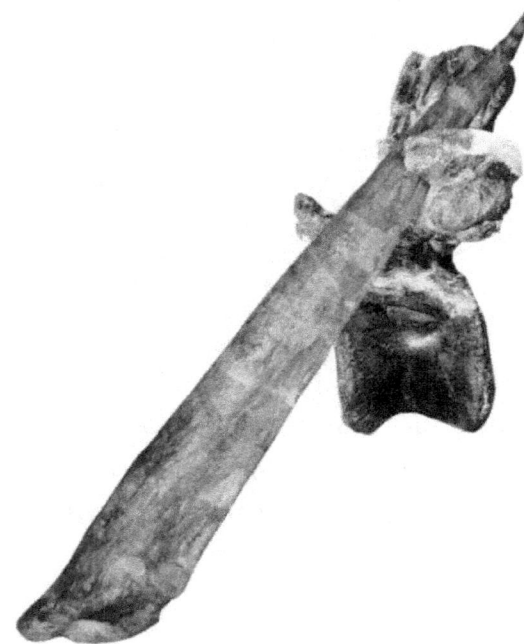

Even the mighty sauropods would have fallen victim to *Allosaurus*. They would likely be used as both live prey and as objects of scavenging. This is based off scrapings on sauropod bones that fit allosaur teeth and the presence of shed allosaur teeth in association with sauropod bones[148].

It is unlikely that *Allosaurus* would have gone after adult sauropods due to being greatly outsized by them unless they hunted in packs[149]. Pack hunting in large

---

[148] Fastovsky & Smith, 2004
[149] Paul, 1988

theropod dinosaurs is not unheard of in the fossil record. After all, evidence from the Dry Island Bonebed in Alberta, Canada, suggests 26 *Albertosaurus* of varying ages living together[150].

But how would Al have brought down his prey? Research in the 1990's and first decade of the 21st century may have found a solution to this puzzle. American palaeontologist Robert T. Bakker compared *Allosaurus* to Cenozoic sabre-toothed carnivorous mammals as he found similar adaptations in both groups. These include a reduction of jaw muscles, an increase in neck muscles, and the ability to open the jaws extremely wide **(Figure 8.c)**. The short teeth in effect became small serrations on a saw-like cutting edge running the length of the upper jaw, which would have been driven into prey. This type of jaw would permit slashing attacks against larger prey, with the goal of weakening the victim[151].

---

[150] Young, 2011
[151] Bakker, 1998

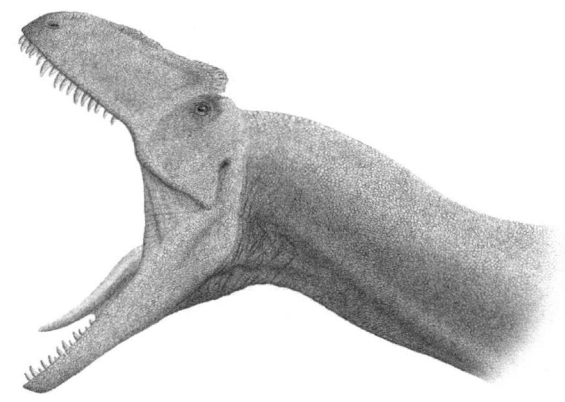

Figure 8.c: *Allosaurus* **attacking, based on the theories of Bakker (1998) and Rayfield *et al.* (2001).**

Evidence gathered from finite element analysis on *Allosaurus* skulls also support Bakker's theory. It found that the skull was very strong and the bite force was very weak. The bite force of *Allosaurus* was only 805 to 2,148 N. The skull, however, could withstand nearly 55,500 N of vertical force against the tooth row[152]. This means that to kill prey, Al would have opened his mouth wide and then use his skull like a hatchet against prey. Slashing at prey with his teeth.

---

[152] Rayfeild *et al*, 2007

Additional evidence to support the 'hatchet' argument comes from a biomechanical study by Ohio University published in 2013. The allosaur specimen used in the study was Al and involved scanning replicas of the skull and neck into a computer and using techniques normally used in engineering to add soft tissues such as muscles to the digitised fossil. It found that *Allosaurus* had an unusually low attachment point on the skull for the longissimus capitis superficialis neck muscle compared to other theropods such as *Tyrannosaurus*. This would have allowed the animal to make rapid and forceful vertical movements with the skull. This suggests a feeding style more like a bird of prey such as a falcon as opposed to the lateroflexive feeding pattern seen in crocodilians and inferred for tyrannosaurs[153].

The shape of the skull meant that *Allosaurus* had binocular vision to 20° of width. This is slightly more than modern crocodiles. This means that Al, like crocodilians, would have been able to judge distances and time

---

[153] Snively *et al*, 2013

attacks[154]. This would have made it a more efficient and effective killer.

Other parts of the body and not just the skull would have been used in hunting and killing prey. The arms, for example, were suited to both grasping hold of prey from a distance and clutching it close[155]. *Allosaurus* could also use its claws to hook onto things due to their articulation[156].

The top speed of *Allosaurus* has been estimated to between 19 to 34 mph[157]. This is much faster than other large theropods. A 2007 study by William Sellers and Phillip Manning suggested that *Tyrannosaurus* could have reached speeds of up to 18 mph and *Allosaurus* 21 mph[158]. More importantly, this was faster than the giant sauropods that Al would have eaten.

Allosaurus may also have fed on each other. One example of cannibalism in *Allosaurus* are shed allosaur teeth found among rib fragments and possible tooth marks

---

[154] Stevens, 2006
[155] Carpenter, 2002
[156] Gilmore, 1920
[157] Christiansen, 1998
[158] Sellers & Manning, 2007

on a shoulder-blade of another *Allosaurus*[159]. Cannibalism in theropods can also be seen in multiple tyrannosaurid fossil specimens[160].

Stress fractures are known in 17 out of 281 foot bones according to a 2001 study by Bruce Rothschild. They concluded that these fractures occurred during interactions with prey as the allosaur tried to hold struggling prey down with its feet. The abundance of stress fractures and avulsion injuries in *Allosaurus* provides evidence for active predatory lifestyles and not scavenging diets[161].

---

[159] Goodchild, 2004
[160] Longrich *et al*, 2010
[161] Rothschild *et al*, 2001

Behaviour

Dinosaurs are often, incorrectly, shown as stupid in the popular media. However, theropod dinosaurs were the most intelligent animals on the planet during the Mesozoic. Many species were more intelligent than Mesozoic mammals[162]. It was this that helped to keep dinosaurs on top.

Endocasts of *Allosaurus* skulls can help show how Al would have behaved. Birds have a large region dedicated to processing information and a small area for sensory inputs. Crocodilian brains have a large area for sensory information and a small area for processing that information. The brain of *Allosaurus* looks more like the crocodilian brain than the bird brain **(Figure 9.a)**. This means that Al would have behaved like a crocodile[163]. However, his skeletal morphology was much more like a bird.

---

[162] Bakker, 2008
[163] Haines, 2001

The structure of the inner ear of *Allosaurus* was like that of a crocodilian. This means that Al could probably have heard lower frequencies best, and would have had trouble with subtle sounds. The olfactory bulbs were large and seem to be well suited for detecting odours, although the area for evaluating smells was relatively small[164].

Figure 9.a: the endocast of *Allosaurus* looks more like that of an alligator than the bird. Image courtesy of Heinrich Mallison.

Allosaurus might also have engaged in advanced social behaviours such as pack hunting. Robert Bakker has

---

[164] Rogers, 1999

interpreted shed *Allosaurus* teeth and chewed bones of large herbivores as evidence of parental care. So maybe the adult allosaurs brought food to lairs for their young to eat until they were grown. This also helped prevent other carnivores from scavenging on the food[165].

Head-biting is common in theropods. An example is *Allosaurus* bite wounds to the skull. The specimen in question is a left dentary from the holotype of *Labrosaurus ferox* which a synonym of *Allosaurus* **(Figure 9.b)**. This injury could have arisen from a number of scenarios. Reasons include; territorial disputes, courtship displays, playing and intrapack dominance[166].

---

[165] Bakker, 1997
[166] Tanke and Currie, 1998

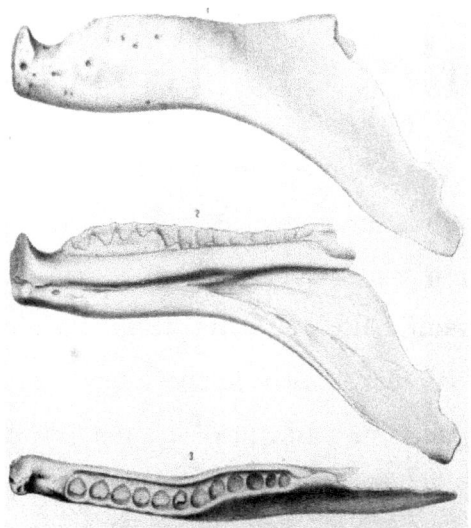

Figure 9.b: The left dentary of *Allosaurus*.

Paleopathology

Arguably the most important, and interesting, aspect of Big Al are his injuries. He has gained international recognition in the paleontological community for his tragically painful life and early death. Paleopathologies are known on other *Allosaurus* specimens. For example, a 2001 study on stress fractures and tendon avulsions in theropods found that 3 out of the 47 hand bones had stress fractures and 17 out of 281 foot bones exhibited stress fractures[167].

However, Al is different. This dinosaur had at least 19 injuries[168] that left their mark on his skeleton. This means there would have been many more that we do not know about because we do not have any of the soft tissue. This high number of injuries may suggest that Al was more aggressive than other allosaurs or he was just clumsy[169].

---

[167] Rothschild *et al*, 2001
[168] Hanna, 2002
[169] Haines, 2001

The manual phalanx II from his right hand displayed evidence of a healed fracture which was surrounded by substantial bone growth **(Figure 10.a)**. A look at a thin section of the bone under a Scanning Electron Microscope revealed an oblique longitudinal healed fracture line through the bone. The injury was likely the result of a twisting motion on the digit[170].

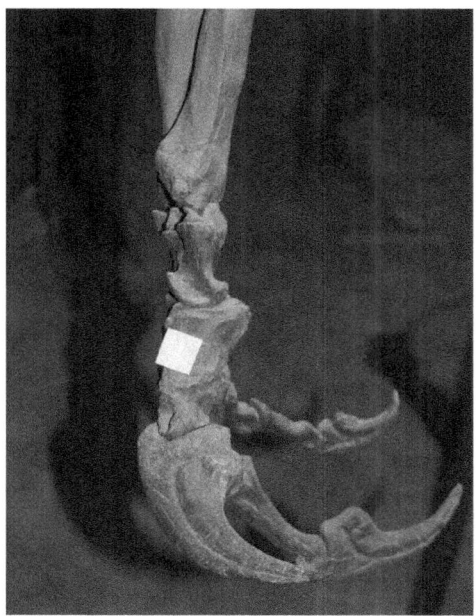

Figure 10.a: The manual phalanx II (with marker) showing extra bone growth covering a fracture. Image courtesy of Laura Vietti.

---

[170] Rega, 2012

He received an injury to the second vertebrae in his tail 1 or 2 years before he died. The chevron was broken in half and it healed back together[171]. New bone growth is called exostosis and is present on the third dorsal vertebra. However when dealing with bone abnormalities you should not always presume that the regrowth was the result of an injury as the third left dorsal rib shows no sign of trauma[172].

Al's ribs were also prone to injury **(Figure 10.b & c)**. The third and fifth dorsal ribs show a hard rugose callus formed by bone regrowth. The third, fourth and fifth dorsal ribs have traumatic healed fractures with callus formation which resulted from a violent impact from the right side [173]. The third ribs also show evidence of misalignment during healing. The fifth shows a putative cloaca for pus drainage. The bone infection may have been formed as a result of complications during the fractures healing[174].

---

[171] Haines, 2001
[172] Hanna, 2002
[173] Hanna, 2002
[174] Rega, 2012

The sixth dorsal rib has a 6.5 cm long projection on the bone called a spicule on the lateral side. Additionally, the fourteenth right dorsal rib has two bone spicules. The origins for these particular pathologies are unclear. The one on the sixth rib does not look like a healed fracture so might be developmental. The fourteenth rib may be either traumatic, developmental or isopathic[175].

Figure 10.b: Some of Al's ribs which have pathologies. Image courtesy of Laura Vietti.

---

[175] Hanna, 2002

Figure 10.c: Some of Al's ribs which have pathologies. Image courtesy of Laura Vietti.

Al would also go on to hurt his backbone, shoulder and pelvis[176]. Additional pathologies can be seen on the sixth neck vertebra, the third eighth and thirteenth back vertebrae, the gastralia, right scapula, manual phalanx I,

---

[176] Haines, 2001

left ilium, metatarsals III and V **(Figure 10.d)**, the first phalanx of the third toe and third phalanx of the second. The ilium had "a large hole... caused by a blow from above"[177].

Figure 10.d: Al's metatarsal V showing abnormal growth due to injury. Image courtesy of Laura Vietti.

Al's most devastating injury occurred in his right foot on the first phalanx of the third toe **(Figure 10.e & f)**. It was afflicted by an involucrum[178], a new layer of bone

---

[177] Molnar, 2001
[178] Molnar, 2001

outside the existing bone. This resulted from the stripping off of the periosteum by the accumulation of pus within the bone, and new bone growing from the periosteum.

The infection itself was long-lived. Al would have lived with the infection for up to 6 months to a year. The injury would have been very painful and the poor dinosaur would have had to limp. The thickness of the infected toe would have caused it to rub against the other two toes[179]. This injury would have spelt the beginning of the end for Al.

In total, approximately 2% of the bones show some sort of pathology. One-third of them are fractures in various stages of re-healing, including five ribs[180].

---

[179] Haines, 2001
[180] Rega, 2012

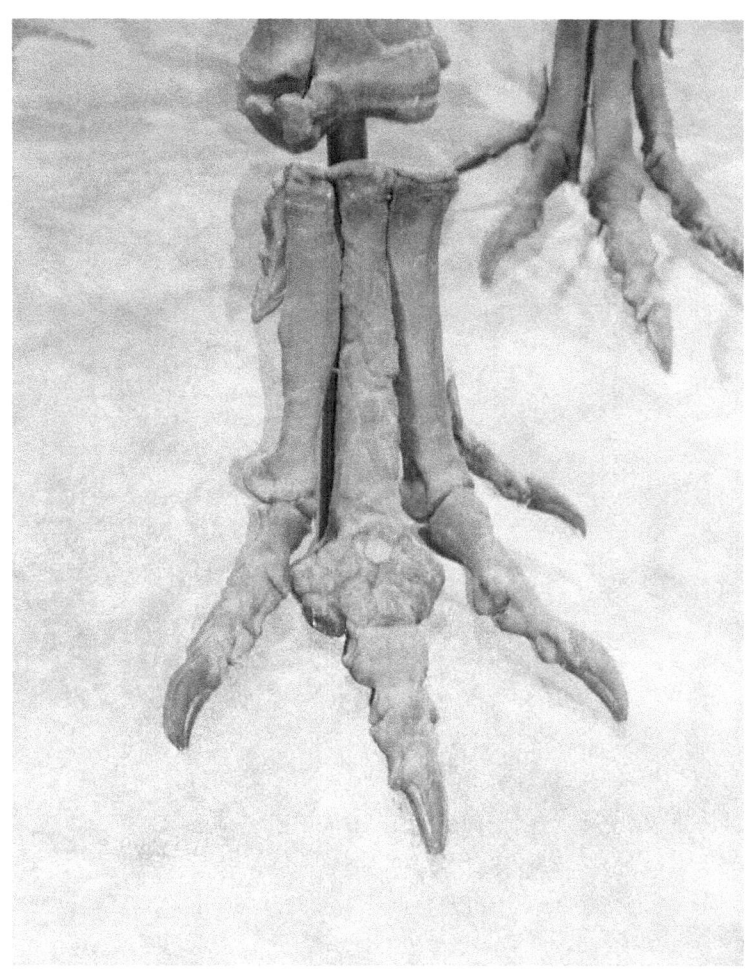

Figure 10.e: Al's infected toe. Image courtesy of Laura Vietti.

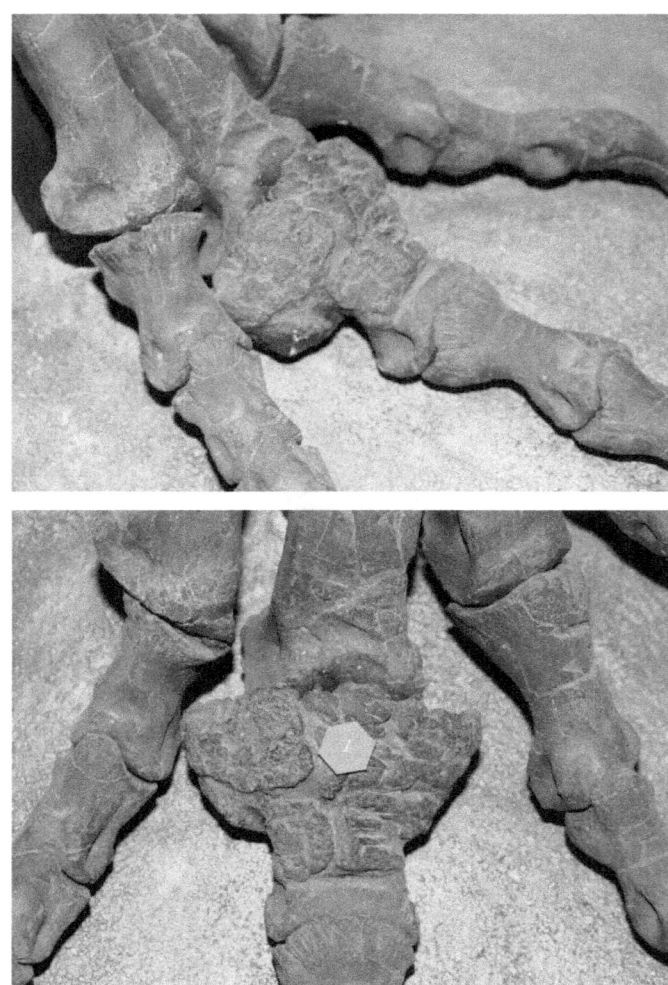

Figure 10.f: Close ups of Al's infected toe. Image courtesy of Laura Vietti.

Death of the dinosaur

With most dinosaurs their final hours are often the easiest to piece together and the same is true for Al. The infection in the toe would have incapacitated Al to such a degree that he would have stopped hunting. He would have struggled to find food and water so his condition would have worsened[181].

Al would have scavenged for food, possibly from kills from other large theropods or dinosaurs that were overcome by the drought that would later hit. If a drought was to hit it would have been enough to tip him over the edge. This is exactly what happened. The drought is evidenced by the fact that Al collapsed in a dried out riverbed[182] and that the Morrison was seasonal with distinct wet and dry seasons[183].

---

[181] Breithaupt, 2001
[182] Idbid
[183] Russell, 1989

Al might have come to the dried out river bed because maybe he knew that this was a source of water and somewhere he would have felt safe. Al could have remained here, hoping that the rains would come. They eventually would, but it was too late.

Overcome by his injuries and hunger Al collapsed in a dry river bed. 95% of the skeleton was present indicating that the body wasn't transported far from where he died[184].

As Al lay in the dry river bed, his carcass desiccated in the hot sun. The muscles, tendons, and other soft tissue dried out, and the animal's head was pulled back over its tail in a characteristic "death pose" **(Figure 11.a)** This is often seen in articulated fossil vertebrates from arid environment [185] **(Figure 11.b)**. Palaeontologists have estimated that Al lay out in the sun for a period of a couple of months[186].

---

[184] Breithaupt, 2001
[185] Idbid
[186] Haines, 2001

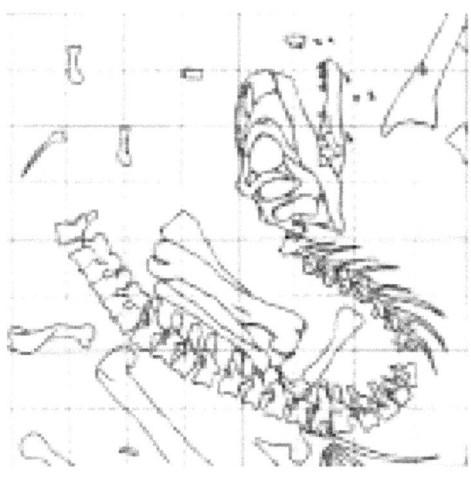

Figure 11.a: A plan showing the position of Al's bones during the excavation. Notice how the spine has been curved back. Image courtesy of Brent Breithaupt.

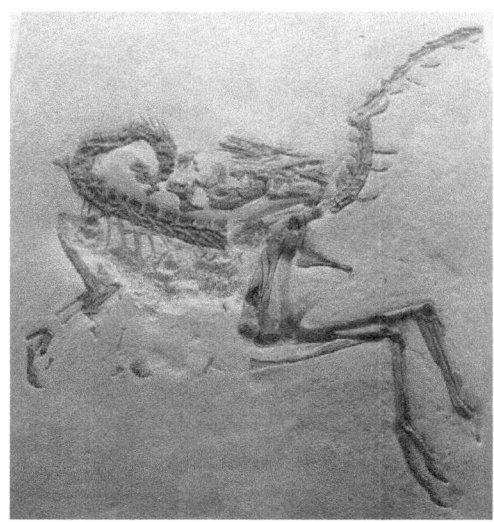

Figure 13.b: Skeleton of *Compsognathus* from the Solnholfen Limestone of Southern Germany. Notice how, like in the Big Al plan above, the spine is arching back.

There is little evidence of vertebrate scavenging on Al's bones. However, the bones do show evidence of beetle burrows, suggesting that the carcass was the dining spot of hundreds to thousands of beetle larvae[187]. This type of damage to the bones is called pseudopathology. The beetle damage was caused after post-mortem exposure, but before the bones were covered by a series of flooding events[188].

Soon, however, the rains arrived and Al's body was covered by sediment. The flood waters that buried the body disarticulated it to some degree. For example, the right ilium was disassociated from the hip region and transported several metres. Many of the ribs and gastralia and other smaller bones (e.g. hand and foot bones) were found disarticulated and scattered around the body. However, the femur, tibia and fibula remained together as the ligaments were not completely decayed.

The specimen was found lying on its left side with its ventral portion tilted upward slightly. This was probably

---

[187] Breithaupt, 2001
[188] Rega, 2012

caused by bloating of the body cavity. This body position allowed for the ischia and associated pubis to be disarticulated from the hip region and moved anteriorly along the axial of the skeleton. It also meant the left side of the skull was well embedded in the sediment and the right side was exposed longer. This is evidenced by a scattering of maxillary teeth and the disarticulation of the lower jaw[189].

---

[189] Breithaupt, 2001

Excavation and Big Al 2

Big Al (MOR 693) was discovered in 1991 in the Upper Jurassic Morrison Formation of Big Horn County, Wyoming[190]. The skeleton was uncovered after a Swiss fossil collecting company, who in 1990, reopened a site discovered by Barnum Brown in 1932. In 1991, they founded a new quarry which would later yield Big Al. Big Al could have been lost to science forever but the Bureau of Land Management, whilst flying over to check for wildfires, noticed the quarries. After checking land records, it soon became clear that the quarry was on Federal land. The Swiss team had permission to collect on the privately owned land, but not the Federal land.

Because the quarry was found on Federal land meant the fossils belonged to the people of the United States and thus managed by the American government. This means

---

[190] Hanna, 2002

that Al is considered a public resource and so cannot be sold into a private collection.

Palaeontologists were brought in from Montana State University's Museum of the Rockies (MOR), the University of Wyoming (UW) Geological Museum and the Royal Tyrrell Museum in Alberta, Canada to survey the site. Al's excavation was one of the most heavily publicised in history with thousands of people going to the quarry to see one of the most complete and important dinosaur skeletons of all time get unearthed[191]. Big Al is now housed at the University of Wyoming Geological Museum **(Figure 12.a)**.

---

[191] Breithaupt, 2001

Figure 12.a: Big Al taking pride of place at the University of Wyoming Geological Museum. Image courtesy of Laura Vietti.

Like with many dinosaur fossils, Al was jacketed to protect the specimen during transportation **(Figure 12.b)**. The process of jacketing involves encasing the bones in a plaster and burlap casing[192].

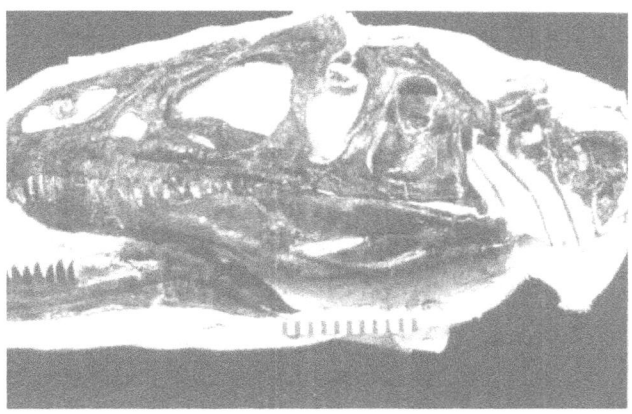

Figure 12.b: The left side of "Big Al" skull as it was being prepared out from the surrounding rock while still in its protective plaster jacket. Image courtesy of Thomas Adams, Pat Leiggi and Brent Breithaupt.

Later in 1996, the same team would discover another *Allosaurus* from the same quarry and named it

---

[192] American Museum of Natural History, 2015

Big Al Two. This new discovery is the most complete example of *Allosaurus* ever discovered. Big Al Two is even more complete than the original Big Al. Like its namesake, Big Al Two exhibits many pathologies[193] **(Figure 12.c)**. Big Al Two is bigger than Big Al at 8.2 metres. Big Al Two may represent a new, undescribed species due to unique skull features. Unlike Big Al, Big Al Two was preserved with evidence of its last meal; a small herbivorous dinosaur ischium, a lungfish tooth, a gastrolith, and some bone fragments[194].

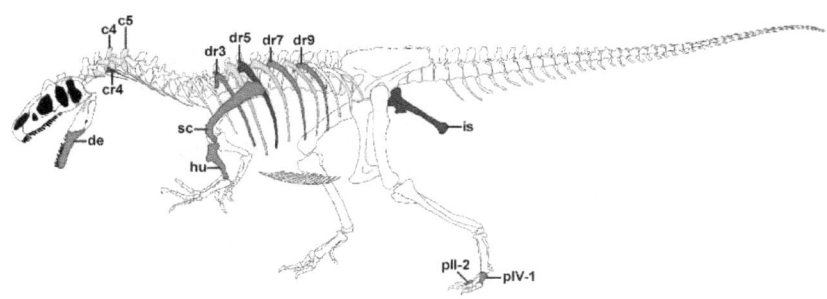

Figure 12.c: Skeletal restoration of Big Al Two showing numerous pathologies.

---

[193] Foth *el al*, 2015
[194] BHI, 2015

Conclusion

Al is special. Not just because of the completeness of the skeleton, but the fossils tell a story. Thanks to his injuries and high degree of preservation we know more about Big Al than just about any other dinosaur. The story of this dinosaur has been pieced together by many scientists, many of whom have been referenced throughout this book, since 1991. In short Al represents a moment in prehistory. A snapshot in the life of a carnivorous dinosaur.

References

American Museum of Natural History. (2015). *Techniques in the Field.* Retrieved from http://preparation.paleo.amnh.org/35/techniques-in-the-field.

Araujo, R. (2013). Filling the gaps of dinosaur eggshell phylogeny: Late Jurassic Theropod clutch with embryos from Portugal. *Scientific Reports* **3**.

Asara, J.M; Schweitzer M.H; Freimark L.M; Phillips, M; Cantley L.C. (2007). Protein Sequences from Mastodon and Tyrannosaurus Rex Revealed by Mass Spectrometry. *Science* **316**(5822): 280–5.

Bakker, R. (1972). Anatomical and ecological evidence of endothermy in dinosaurs. *Nature* **238**(5359): 81–85.

Bakker, R. (1975). Dinosaur Renaissance: The dinosaurs were not obsolescent reptiles but were a novel group of "warm-blooded" animals. And the birds are their descendants. *Scientific America* **232** (4): 58-77.

Bakker, R. (1997). Raptor Family values: Allosaur parents brought giant carcasses into their lair to feed their young. In D.L. Wolberg., E. Sump., G.D. Rosenberg.

(eds.). Dinofest International. *Proceedings of a Symposium Held at Arizona State University.* Philadelphia: *Academy of Natural Sciences.* pp. 51–63.

Bakker, R. (1998). Brontosaur killers: Late Jurassic allosaurids as sabre-tooth cat analogues.

Bakker, R. (2008). Hunting Dinosaurs with Dr. Robert Bakker.
Bakker, R., Bir, G. (2004). Dinosaur Crime Scene Investigations. In P. Currie., E. Koppelhus., M. Shugar., J.Wright (Eds.). *Feathered Dragons.* Bloomington: Indiana University Press.

Bates, K *et al*. (2009). How Big Was 'Big Al'? Quantifying the effect of soft tissue and osteological unknowns on mass predictions for Allosaurus (Dinosauria:Theropoda). *Palaeontologia Electronica* **12** (3).

Benson, R.B.J., Carrano, M.T., Brusatte, S.L. (2010). A new clade of archaic large-bodied predatory dinosaurs (Theropoda: Allosauroidea) that survived to the latest Mesozoic. *Naturwissenschaften* **97**(1): 71–78.

Benton, M (2014). *Vertebrate Palaeontology* (4th ed). London: Wiley-Blackwell.

Benton, M; Harper, D. (2009). *Introduction to Paleobiology and the Fossil Record* (2nd ed). London: Wiley-Blackwell.

BHI. (2015). *Allosaurus sp. BIG AL TWO Skeleton- Fossil Replica.* Retrieved from http://www.bhigr.com/store/product.php?productid=474.

Big Al Uncovered. (2001a). Documentary. Directed by Haines, T. London: BBC.

Breithaupt, B. (2001). *The Case of Big AL the Allosaurus: A Study of Paleodetective Partnerships.* Retrieved from http://www2.nature.nps.gov/geology/paleontology/pub/fossil_conference_6/breithaupt.htm.

Brusatte, S.L., Chure, D.J., Benson, R.B.L., Xu, X. (2010). The osteology of Shaochilong maortuensis, a carcharodontosaurid (Dinosauria: Theropoda) from the Late Cretaceous of Asia. *Zootaxa* **2334**: 1–46.

Butler, J., Porro, L., Galton, P., Chiappe, L. (2012). Anatomy and Cranial Functional Morphology of the Small-Bodied Dinosaur *Fruitadens haagarorum* from the Upper Jurassic of the USA. *PLoS ONE.* **7**(4): 1-31.

Carpenter, K. (2002). Forelimb biomechanics of nonavian theropod dinosaurs in predation. *Senckenbergiana Lethaea* **82**(1): 59–76.

Carpenter, K., Sanders, F., McWhinney, L., Wood, L. (2005). Evidence for predator-prey relationships: Examples for *Allosaurus* and *Stegosaurus*". In K. Carpenter (ed.). *The Carnivorous Dinosaurs* (pp. 325-350. Bloomington: Indiana University Press.

Carrano, M. T., Benson, R. B. J., Sampson, S. D. (2012). The phylogeny of Tetanurae (Dinosauria: Theropoda). *Journal of Systematic Palaeontology* **10**(2): 211–300.

Carrano, M., Mateus O., & Mitchell J. (2013). First definitive association between embryonic *Allosaurus* bones and prismatoolithus eggs in the Morrison Formation (Upper Jurassic, Wyoming, USA). *Annual Meeting of Vertebrate Paleontology*. **101**: *Journal of Vertebrate Paleontology, Program and Abstracts, 2013*.

Chiappe, L. M. (2009). Downsized Dinosaurs: The Evolutionary Transition to Modern Birds. *Evolution: Education and Outreach* **2**(2): 248–256.

Christiansen, P. (1998). Strength indicator values of theropod long bones, with comments on limb proportions and cursorial potential. *Gaia* **15**: 241–255.

Chure, D., Madsen, J. (1996). On the presence of furculae in some non-Maniraptorian theropods. *Journal of Vertebrate Paleontology* **16** (3): 573-577.

Clarkson, E.N.K. (1993). *Invertebrate Palaeontology and Evolution* (3rd ed). London: Chapman & Hall.

Coria, R.A., Salgado, L. (1995). A new giant carnivorous dinosaur from the Cretaceous of Patagonia. *Nature* **377**: 225-226.

Currie, J. (2005). *Dinosaur Provincial Park: A Spectacular Ancient Ecosystem Revealed.* Bloomington: Indiana University Press.

Currie, J., Hurum, J., Sabath, K. (2003). Skull structure and evolution in tyrannosaurid dinosaurs. *Acta Palaeontologica Polonica* **48** (2): 227-234.

Currie, P.J., Azuma, Y. (2006). New specimens, including a growth series, of Fukuiraptor (Dinosauria, Theropoda) from the Lower Cretaceous Kitadani Quarry of Japan. *Journal of the Paleontological Society of Korea* **22**(1): 173-193.

Donald, G. (1997). *Dinosaurs: The Encyclopedia.* Jefferson: McFarland & Co.

Eddy, D.R., Clarke, J.A. (2011). New Information on the Cranial Anatomy of *Acrocanthosaurus atokensis* and Its Implications for the Phylogeny of Allosauroidea (Dinosauria: Theropoda). *PLOS one* **6**(3): e17932.

Erickson, G., Makovicky, P., Currie, P., Norell., Yerby, S., Brochu, C. (2004). Gigantism and comparative life-history parameters of tyrannosaurid dinosaurs. *Nature* **430**(7001): 772–775.

Ezcurra, M., Agnolín, F. (2012). A New Global Palaeobiogeographical Model for the Late Mesozoic and Early Tertiary. *Systematic Biology*. **61**(4): 553-566.

Fastovsky, D., Smith, J. (2004). Dinosaur Paleoecology. In D.B. Weishampel., P. Dodson., H. Osmólska (Eds.). *The Dinosauria*. Oakland: University of California Press.

Fernandes de Azevedo, R.P., Simbras, F.M., Futado, M.R., Candeiro, C.R.A., Bergqvist, L.P. (2013). First Brazilian carcharodontosaurid and other new theropod dinosaur fossils from the Campanian-Maastrichtian Presidente Prudente Formation, São Paulo State, southeastern Brazil. *Cretaceous Research* **40**: 131-142.

Foos, A., Hannibal, J. (1999). *Geology of the Dinosaur National Monument*. Retrieved from http://nature.nps.gov/geology/education/Foos/dino.pdf.

Foster, J. (2007). *Jurassic West: The Dinosaurs of the Morrison Formation and Their World.* Bloomington: Indiana University Press.

Foster, J. (2009). Preliminary body mass estimates for mammalian genera of the Morrison Formation (Upper Jurassic, North America). *PaleoBios* **28**(3):114-122.

Foth, C., Evers, S., Pabst, B., Mateus, O., Flisch, A., Patthey, M., Rauhut., O.W.M. (2015). New insights into the lifestyle of *Allosaurus* (Dinosauria: Theropoda) based on another specimen with multiple pathologies. *PeerJ PrePrints* **3**: e824v1.

Gates, T. (2005). The Late Jurassic Cleveland-Lloyd Dinosaur Quarry as a Drought-Induced Assemblage. *Palaios* **20**: 363-375.

Geoweb. (2015). *The Reality of Biostratigraphy and the correlation of the Morrison Formation.* Retrieved from http://geoweb.gg.uwyo.edu/geol2100/reality%20of%20Biostratigraphy_Nov17.pdf.

Gilmore, C. (1920). Osteology of the carnivorous Dinosauria in the United States National Museum, with special reference to the genera *Antrodemus* (*Allosaurus*) and Ceratosaurus. *Bulletin of the United States National Museum* **110**: 1–159.

Goodchild, D.B. (2004). A new specimen of *Allosaurus* from north-central Wyoming. *Journal of Vertebrate Paleontology* **24** (3, Suppl.): 65A.

Haines, T (1999). *Walking with Dinosaurs: A Natural History*. London: BBC Worldwide.

Hanna, R. (2002). Multiple Injury and Infection in a Sub-Adult Theropod Dinosaur *Allosaurus fragilis* with Comparisons to Allosaur Pathology in the Cleveland-Lloyd Dinosaur Quarry Collection. *Journal of Vertebrate Palaeontology* **22**(1): 76-90.

Harris, J.D., Dodson, P. (2004). A new diplodocoid sauropod dinosaur from the Upper Jurassic Morrison Formation of Montana, USA. *Acta Palaeontologica Polonica* **49**(2): 197–210.

Hendrickx, C., Mateus, O.V. (2014). *Torvosaurus* gurneyi n. sp., the Largest Terrestrial Predator from Europe, and a Proposed Terminology of the Maxilla Anatomy in Nonavian Theropods. *PLoS ONE* **9**(3).

Home, D. (2016). *The Tyrannosaur Chronicles: The Biology of the Tyrant Dinosaurs.* New York: Bloomsbury

Hummel, J., Gee, C.T., Südekum, K.H., Sander, P.M., Nogge, G., Clauss, M. (2008). In vitro digestibility of fern and gymnosperm foliage: implications for sauropod feeding ecology and diet selection. *Proceedings of the Royal Society B* **275**:1015-1021.

Ji, Q., Ji, S. (1997). A Chinese archaeopterygian, Protarchaeopteryx gen. nov. *Geological Science and Technology (Di Zhi Ke Ji)*, **238**: 38-41.

Ji, Q., Ji, S. (1996). On discovery of the earliest bird fossil in China and the origin of birds. *Chinese Geology* **10**(233): 30–33.

Klappenback, L. (2014). How Many Animal Species Inhabit Our Planet. Retreved from http://animals.about.com/od/zoologybasics/a/howmanyspecies.htm.

Koeberl, C., MacLeod, K. (2002). *Catastrophic Events and Mass Extinctions: Impacts and Beyond.* U.S.A: Geological Society of America.

Larson, P. (1998). The Theropod Reproductive System. *Gaia* **15**: 389-397.

Lee, A., Werning, S. (2008). Sexual maturity in growing dinosaurs does not fit reptilian growth models. *Proceedings of the National Academy of Sciences of the United States of Am*erica **105**(2): 582–587.

Longrich, N., Horner, J., Erickson, G., Currie, P. (2010). Cannibalism in *Tyrannosaurus rex*. *PLOS ONE* **5**(10): e13419.

Marsh, O. (1878). Notice of new dinosaurian reptiles. *American Journal of Science and Arts* **15**: 241–244.

Martill, D.M. (1988). *Leedsichthys problematicus*, a giant filter-feeding teleost from the Jurassic of England and France. *Neues Jahrbuch fur Geologie und Palaontologie* Monatshefte **1988**(11): 670-680.

McWhinney, L., Rothschild, B., Carpenter, K. (2001). Posttraumatic Chronic Osteomyelitis in Stegosaurus dermal spikes. In K. Carpenter, (Ed). *The Armored Dinosaurs*. Indiana University Press. pp. 141–56.
Middleton, K., Gatesy, S. (2000). Theropod forelimb design and evolution. *Zoological Journal of the Linnean Society* **128**(2): 149–187.

Molnar, R. E. (2001,). Theropod paleopathology: a literature survey. In D.H Tanke., K. Carpenter (Eds.).

Mesozoic Vertebrate Life pp. 337-363). Bloomington: Indiana University Press.

Motani, R. (2000). *Rulers of the Jurassic Seas.* Retrieved from http://www.karencarr.com/News/motani/1200 motani.html.

Naish, D., Martill, D.M. (2007). Dinosaurs of Great Britain and the role of the Geological Society of London in their discovery: basal Dinosauria and Saurischia. *Quarterly Journal of the Geological Society* **164**: 493–510.

National Park Service. (2014). *Morrison Formation.* Retrieved from http://www.nps.gov/dino/naturescience/morris on-formation.htm.

National Park Service. (2015). *Dinosaur National Monument Visitation Statistics.* Retrieved from: http://www.nps.gov/dino/learn/management/statistics.htm.

Neovenator salerii. (2016). Retrieved from Dinosaur Isle website: http://www.dinosaurisle.com/neovenator.aspx.

Noe, L., Liston, J., Evans, M. (2003). The first relatively complete exoccipital-opisthotic from the braincase of the

Callovian pliosaur, *Liopleurodon*. *Geological Magazine* **140**(4): 479–486.

Norell, M.A., Clark, J. M., Dashzeveg, D., Barsbold, T., Chiappe, L.M., Davidson, A.R., McKenna, M.C., Novacek, M.J. (1994). A theropod dinosaur embryo and the affinities of the Flaming Cliffs Dinosaur eggs. *Science* **266**(5186): 779–82.

Novas, F.E., Agnolín, F.L., Ezcurra, M.D., Canale, J.I., Porfiri, J.D. (2012). Megaraptorans as members of an unexpected evolutionary radiation of tyrant-reptiles in Gondwana. *Ameghiniana* **49**(Suppl.): R33.

Novas, F.E., Ezcurra, M.D., Lecuona, A. (2008). *Orkoraptor burkei* nov. gen. et sp., a large theropod from the Maastrichtian Pari Aike Formation, Southern Patagonia, Argentina. *Cretaceous Research* **29**(3): 468–480.

Paul, G.S. (1988). *Predatory Dinosaurs of the World.* New York: Simon & Schuster.

Paul, G.S. (2010). *Dinosaurs: A Field Guide.* London: A&C Black.

Perez-Moreno, B. P., Chure, D.J., Pires, C., Marques da Silva, C., Santos, V.D., Dantas, P., Povoas, L., Cachão,

M., Sanz, J.L., Galopim De Carvalho, A.M. (1999). On the presence of *Allosaurus fragilis* (Theropoda: Carnosauria) in the Upper Jurassic of Portugal: first evidence of an intercontinental dinosaur species. *Journal of the Geological Society* **156**(3): 449-452.

Peterson, I. (2002). *Whips and Dinosaur Tails*. Retreived from https://www.sciencenews.org/article/whipsand-dinosaur-tails.

Rauhut, O.W.M. (2011). Theropod dinosaurs from the Late Jurassic of Tendaguru (Tanzania). *Special Papers in Palaeontology* **86**:195-239

Rayfield, E. (2007). Finite element analysis in vertebrate morphology. *Annual Reviews in Earth and Planetary Sciences* **35**: 541–576.

Rega, E. (2012). Disease in Dinosaurs. In M. Brett-Surman., T. Holtz., J. Farlow (Eds.). *The Complete Dinosaur* (pp. 667-713). Bloomington: Indiana University Press.

Rogers, R., Eberth, D., Fiorillo, A. (2007). *Bonebeds: Genesis, Analysis, and Paleobiological Significance*. Chicago: The University of Chicago Press.

Rogers, S. (1999). *Allosaurus*, crocodiles, and birds: Evolutionary clues from spiral computed tomography of an endocast. The *Anatomical R*ecord **257**(5): 163–173.

Rothschild, B., Tanke, D. H., and Ford, T. L. (2001). Theropod stress fractures and tendon avulsions as a clue to activity. In D.H. Tanke., K. Carpenter (Eds.). *Mesozoic Vertebrate Life* (pp. 331-336). Bloomington: Indiana University Press.

Rowe, T., Gauthier, J. (1990). Ceratosauria. In D.B. Weishampel., P. Dodson., H. Osmólska. The *Din*osauria. Oakland: University of California Press.

Russell, Dale A. (1989). *An Odyssey in Time: Dinosaurs of North Americ*a. Minocqua, Wisconsin: NorthWord Press.

Sampson, S. (2009). *Dinosaur Odyssey: Fossil Threads into the Web of Time*. Los Angeles: University of California Press.

Schweitzer, M., Wittmeyer, J.; Horner, J.; Toporski, J. (2005). Soft-tissue vessels and cellular preservation in *Tyrannosaurus rex*. *Science* **307**(5717): 1952–5.

Selden, P., Nudds, J. (2012). *Evolution of Fossil Ecosystems* (2nd ed). London: Manson Publishing.

Sellers, W.L., Manning, P.L. (2007). Estimating dinosaur maximum running speeds using evolutionary robotics. *Proc. R. Soc. B* **274** (1626).

Sereno, P., Martinez, R., Wilson, J., Varricchio, D., Alcober, O., Larsson, H. (2008). Evidence for Avian Intrathoracic Air Sacs in a New Predatory Dinosaur from Argentina. *PLoS ONE* **3**(9): e3303.

Smith, D. (1998). A morphometric analysis of *Allosaurus*. *Journal of Vertebrate Paleontology* **18**(1): 126–142.

Smith, J. B., Vann, D.R., Dodson, P. (2005). Dental morphology and variation in theropod dinosaurs: Implications for the taxonomic identification of isolated teeth. *The Anatomical Record Part A: Discoveries in Molecular, Celluar, and Evolutionary Biology* **285A**(2): 699-736.

Smithsonian. (2014). *Jurassic Climate and Tectonic Activity.* Retrieved from http://www.paleobiology.si.edu/geotime/main/htmlVersion/jurassic5.html.

Snively, E., Cotton, J., Ridgely, R., Witmer, Lawrence M. (2013). Multibody dynamics model of head and neck function in *Allosaurus* (Dinosauria, Theropoda). *Palaeontologica Electronica* **16**(2).

Spinar, Z. (1995). *Life Before Man* (5th ed). London: Thames and Hudson.

Stevens, K. (2006). Binocular vision in theropod dinosaurs. *Journal of Vertebrate Paleontology* **26**(2): 321–330.

Stevens, K., Parrish, J. (1999). Neck posture and feeding habits of two Jurassic sauropod dinosaurs. *Science* **284**(5415): 798–800.

Strauss, B. (2014). *10 Facts About Allosaurus.* Retrieved from http://dinosaurs.about.com/od/dinosaurbasics/a/allosaurusfacts.htm.

Sullivan, R. (2014). *Dinosaur Blood Ran Just Right: Not Warm, Not Cold.* Retrieved from http://news.discovery.com/animals/dinosaurs/d inosaur-blood-ran-just-right-not-warm-not-cold140613.htm.

Sun, G., Dilcher, D., Zheng, S., Zhou, Z. (1998). In Search of the First Flower: A Jurassic Angiosperm, *Archaefructus*, from Northeast China. *Science* **282**(5394). 1692-1695.

Switek, B. (2013). *Ancient Fish Downsized But Still Largest Ever*. Retrieved from http://news.nationalgeographic.com/news/2013/08/130827-paleontology-leedsichthys-problematicus-fish-oceans-science/.

Tang, C. (2014). *Morrison Formation*. Retrieved from http://www.britannica.com/EBchecked/topic/393013/Morrison-Formation.

Tanke, D.H., Currie, J. (1998). Head-biting behaviour in theropod dinosaurs: paleopathological evidence. *Gaia* **15**: 167-184.

Taylor, T.N., Taylor, E.L., Krings, M. (2009). *Paleobotany: The biology and evolution of fossil plants* (2nd ed). Academic Press: Cambridge, MA.

Therrien, F., Henderson, D.M. (2007). My theropod is bigger than yours...or not: estimating body size from skull length in theropods. *Journal of Vertebrate Palaeontology* **27**(1): 108-115.

Tidwell, V., Carpenter, K. & Meyer, S. (2001). New Titanosauriform (Sauropoda) from the Poison Strip Member of the Cedar Mountain Formation (Lower Cretaceous), Utah. In D. H. Tanke., K. Carpenter (Eds.).

Mesozoic Vertebrate Life (pp. 139-165). Bloomington: Indiana University Press.

Trujillo, K., Chamberlain, K., Strickland, A. (2006). Oxfordian U/Pb ages from SHRIMP analysis for the Upper Jurassic Morrison Formation of southeastern Wyoming with implications for biostratigraphic correlations. *Geological Society of America Abstracts with Programs* **38**(6): 7.

Turner, A., Makovicky, P., Norell, M. (2012). A review of dromaeosaurid systematics and paravian phylogeny. *Bulletin of the American Museum of Natural History* **371**: 1–206.

Turner, C., Peterson, F., (1999). Biostratigraphy of dinosaurs in the Upper Jurassic Morrison Formation of the Western Interior, U.S.A. pp. 77–114 in Gillette, D. (ed.), *Vertebrate Paleontology in Utah*. Utah Geological Survey Miscellaneous Publication 99-1.

Varela, A.N., Poire, D.G., Martin, T., Gerdes, A., Goin, F.J., Gelfo, J.N., Hoffmann, S. (2012). U-Pb zircon constraints on the age of the Cretaceous Mata Amarilla Formation, Southern Patagonia, Argentina: its relationship with the evolution of the Austral Basin. *Andean Geology* **39**(3): 359-379.

Vermeij, G.J. (1977). The Mesozoic Marine Revolution: Evidence from Snails, Predators and Grazers. *Palaeobiology* **3**: 245–258.

Walker, A.D. (1964). Triassic reptiles from the Elgin area: Ornithosuchus and the origin of carnosaurs. *Philosophical Transactions of the Royal Society of London, Series B, Biological Sciences* **248**: 53–134.

Witton, M. (2013). *Pterosaurs*. Oxford: Princeton University Press.

Wu X., Currie, P.J., Dong Z., Pan S., Wang T. (2009). A new theropod dinosaur from the Middle Jurassic of Lufeng, Yunnan, China. *Acta Geologica Sinica* **83**(1): 9–24.

Xu, X., Clark, J.M., Forster, C.A., Norell, M.A., Erickson, G.M., Eberth, D.A., Jia, C., Zhao, Q. (2006). A basal tyrannosauroid dinosaur from the Late Jurassic of China. *Nature Letters* **439**(9): 715-718.

Xu, X., Norell, M.A., Kuang, X., Wang, X., Zhao, Q; Jia, C. (2004). Basal tyrannosauroids from China and evidence for protofeathers in tyrannosauroids. *Nature* **431**(7009): 680–684.

Yi-Ming, H., Clark, J.M., Xing, X. (2013). A large theropod metatarsal from the upper part of Jurassic Shishugou Formation in Junggar Basin, Xinjiang, China. *Vertrbrata PalAsiatica* **51**(1): 29-42.

Young, J. (2011). *Dino Gangs*. London: Collins.

Zhao, X., Currie, P.J. (1993). A large crested theropod from the Jurassic of Xinjiang, People's Republic of China. *Canadian Journal of Earth Sciences* **30**(10): 2027-2036.

Zhiming, D. (1975). A new carnosaur from Yongchuan County, Sichuan Province. *Ke Xue Tong Bao* **23**(5): 302-304

Image Permissions

Figure 1.a- "Allosaurus size comparison" by Steveoc 86 Marmelad Scott Hartman, [2]. - File:Allosaurus size comparison.svg is an altered version of an image by User:Marmelad which inturn is a vectorized version of an image by User:Dropzink (seen to the right). Licensed under CC BY-SA 2.5 via Wikimedia Commons - http://commons.wikimedia.org/wiki/File:Allosaurus_size_comparison.svg#/media/File:Allosaurus_size_comparison.svg

Figure 1.b- "Allosaurus SDNHM (1)" by Allosaurus_SDNHM.jpg: Sheep81derivative work: Creoqueteamo - This file was derived from: Allosaurus SDNHM.jpg. Licensed under CC BY-SA 3.0 via Wikimedia Commons - http://commons.wikimedia.org/wiki/File:Allosaurus_SDNHM_(1).jpg#/media/File:Allosaurus_SDNHM_(1).jpg

Figure 1.c- "Allosaurus-crane" by Bob Ainsworth http://bobainsworth.com - http://morguefile.com/archive/?display=51268&&MORGUEFILE=n02q52nbjjfak7a2nu4mitq194. Licensed under CC BY 2.0 via Wikimedia Commons - http://commons.wikimedia.org/wiki/File:Allosaurus-crane.jpg#/media/File:Allosaurus-crane.jpg

Figure 1.d- "Allosaurus-fragilis-Klauen" by Domser - Own work. Licensed under CC BY-SA 3.0 via Wikimedia Commons - http://commons.wikimedia.org/wiki/File:Allosaurus-fragilis-Klauen.JPG#/media/File:Allosaurus-fragilis-Klauen.JPG

Figure 1.e- Own work

Figure 1.f- Own work

Figure 1.g- Own work

Figure 1.h- Own work

Figure 1.i- Own work

Figure 2.a- "Cope-and-marsh" by George Bird Grinnell and Marcus Benjamin, respectively - Combination/derivative of Image:OthnielCharlesMarsh1.jpg and Image:Cope Edward Drinker 1840-1897.png. Licensed under Public Domain via Wikimedia Commons - http://commons.wikimedia.org/wiki/File:Cope-and-marsh.png#/media/File:Cope-and-marsh.png

Figure 3.a- Own work

Figure 3.b- Own work

Figure 3.c- Own work

Figure 4.a- "Alloquarry copy" by J. Spencer - Own work. Licensed under Public Domain via Wikimedia Commons - https://commons.wikimedia.org/wiki/File:Alloquarry_copy.png#/media/File:Alloquarry_copy.png

Figure 4.b- "Dinosaur National Monument-inside the Dinosaur Quarry building" by Original uploader was Kevin Saff at en.wikipedia - Transferred from en.wikipedia; originally from http://www.cr.nps.gov/museum/treasures/html/Q/h020.html. Licensed under Public Domain via Wikimedia Commons - https://commons.wikimedia.org/wiki/File:Dinosaur_National_Monument-inside_the_Dinosaur_Quarry_building.jpeg#/media/File:Dinosaur_National_Monument-inside_the_Dinosaur_Quarry_building.jpeg

Figure 4.c- "Green River UT 2005-10-14 2104" by Photo by Michael Overton. - Digital photograph. Licensed under CC BY-SA 2.5 via Wikimedia Commons - https://commons.wikimedia.org/wiki/File:Green_River_UT_2005-10-14_2104.jpg#/media/File:Green_River_UT_2005-10-14_2104.jpg

Figure 5.a- (top) "Saurischia pelvis structure" by Fred the Oyster. Licensed under CC BY-SA 4.0 via Wikimedia

Commons - http://commons.wikimedia.org/wiki/File:Saurischia_pelvis_structure.svg#/media/File:Saurischia_pelvis_structure.svg. (Bottom) "Ornithischia pelvis structure" by Fred the Oyster. Licensed under CC BY-SA 4.0 via Wikimedia Commons - http://commons.wikimedia.org/wiki/File:Ornithischia_pelvis_structure.svg#/media/File:Ornithischia_pelvis_structure.svg

Figure 5.b- "Marsh Camptosaurus" by O.C. Marsh - http://www.copyrightexpired.com/earlyimage/bones/display_marsh_camptosaurus.htm. Licensed under Public Domain via Wikimedia Commons - http://commons.wikimedia.org/wiki/File:Marsh_Camptosaurus.jpg#/media/File:Marsh_Camptosaurus.jpg

Figure 5.c- "Stegosaurus BW" by Nobu Tamura (http://spinops.blogspot.com) - Own work. Licensed under CC BY 2.5 via Wikimedia Commons - http://commons.wikimedia.org/wiki/File:Stegosaurus_BW.jpg#/media/File:Stegosaurus_BW.jpg

Figure 5.d- "Fruitadens" by Smokeybjb - Own work. Licensed under CC BY-SA 3.0 via Wikimedia Commons - saurophhttp://commons.wikimedia.org/wiki/File:Fruitadens.jpg#/media/File:Fruitadens.jpg

Figure 5.e- "Amphicoelias altus scale" by Matt Martyniuk - Own work. Licensed under CC BY 3.0 via Wikimedia Commons - http://commons.wikimedia.org/wiki/File:Amphicoelias_altus_scale.png#/media/File:Amphicoelias_altus_scale.png

Figure 5.f- "SeismosaurusDB" by Original uploader was ДиБгд at ru.wikipedia - Originally from ru.wikipedia; description page is/was here.. Licensed under Public Domain via Wikimedia Commons - http://commons.wikimedia.org/wiki/File:SeismosaurusDB.jpg#/media/File:SeismosaurusDB.jpg

Figure 5.g- "Apatosaurus33" by Original uploader was ДиБгд at ru.wikipedia Anatomical corrections by FunkMonk and Dinoguy2. - Originally from ru.wikipedia; description page is/was here.. Licensed under Public Domain via Wikimedia Commons - http://commons.wikimedia.org/wiki/File:Apatosaurus33.jpg#/media/File:Apatosaurus33.jpg

Figure 5.h- "Brachiosaurus DB" by Богданов dmitrchel@mail.ru - Own work. Licensed under Public Domain via Wikimedia Commons - http://commons.wikimedia.org/wiki/File:Brachiosaurus_DB.jpg#/media/File:Brachiosaurus_DB.jpg

Figure 5.i- "Tanycolagreus topwilsoni" by Smokeybjb - Own work. Licensed under CC BY-SA 3.0 via

Wikimedia Commons - http://commons.wikimedia.org/wiki/File:Tanycolagreus_topwilsoni.jpg#/media/File:Tanycolagreus_topwilsoni.jpg

Figure 6.a- "Earth During the Jurassic Time Period" by Buildingme11 - Own work. Licensed under CC BY-SA 3.0 via Wikimedia Commons - **Error! Hyperlink reference not valid.**

Figure 6.b- "Hamites" by Neale Monks at the English language Wikipedia. Licensed under CC BY-SA 3.0 via Wikimedia Commons - https://commons.wikimedia.org/wiki/File:Hamites.jpg#/media/File:Hamites.jpg

Figure 6.c- "Ichthyosauria offspring in Vienna" by Tommy from Arad - Ichthyosauria offspringsUploaded by FunkMonk. Licensed under CC BY 2.0 via Wikimedia Commons - http://commons.wikimedia.org/wiki/File:Ichthyosauria_offspring_in_Vienna.jpg#/media/File:Ichthyosauria_offspring_in_Vienna.jpg

Figure 6.d- "Pterosaur wing BW2" by Pterosaur_wing_BW.jpg: ArthurWeasley email:aweasley@hotmail.comderivative work: Dinoguy2 (talk) 18:45, 12 March 2009 (UTC) - Pterosaur_wing_BW.jpg. Licensed under CC BY 3.0 via Wikimedia Commons -

https://commons.wikimedia.org/wiki/File:Pterosaur_wing_BW2.jpg#/media/File:Pterosaur_wing_BW2.jpg

Figure 6.e- "Archaeopteryx lithographica (Berlin specimen)" by H. Raab (User: Vesta) - Own work. Licensed under CC BY-SA 3.0 via Wikimedia Commons - https://commons.wikimedia.org/wiki/File:Archaeopteryx_lithographica_(Berlin_specimen).jpg#/media/File:Archaeopteryx_lithographica_(Berlin_specimen).jpg

Figure 7.a- Image courtesy of Ricardo Araújo.

Figure 7.b- "Tyrantgraph". Licensed under Public Domain via Wikimedia Commons - http://commons.wikimedia.org/wiki/File:Tyrantgraph.png#/media/File:Tyrantgraph.png

Figure 8.a- Image courtesy of Kenneth Carpenter

Figure 8.b- Image courtesy of Kenneth Carpenter

Figure 8.c- "Allosaurus Jaws Steveoc86" by I, Steveoc 86. Licensed under CC BY 2.5 via Wikimedia Commons - http://commons.wikimedia.org/wiki/File:Allosaurus_Jaws_Steveoc86.jpg#/media/File:Allosaurus_Jaws_Steveoc86.jpg

Figure 9.a- Image courtesy of Heinrich Mallison

Figure 9.b- "Labrosaurus" by Gilmore, Charles W. - http://archive.org/stream/osteologyofcarni00gilm#page/n205/mode/2up. Licensed under Public Domain via Wikimedia Commons - https://commons.wikimedia.org/wiki/File:Labrosaurus.jpg#/media/File:Labrosaurus.jpg

Figure 10.a- Image courtesy of Laura Vietti

Figure 10.b- Images courtesy of Laura Vietti

Figure 10.c- Image courtesy of Laura Vietti

Figure 10.d- Image courtesy of Laura Vietti

Figure 10.e- Images courtesy of Laura Vietti

Figure 12.b- Image courtesy of Thomas Adams, Brent Breithaupt and Pat Leiggi

Figure 11.b- "Compsognathus longipes cast2". Licensed under CC BY-SA 3.0 via Wikimedia Commons - http://commons.wikimedia.org/wiki/File:Compsognathus_longipes_cast2.jpg#/media/File:Compsognathus_longipes_cast2.jpg

Figure 12.a- Image courtesy of Laura Vietti

Figure 12.b- Image Courtesy of Thomas Adams, Brent Breithaupt and Pat Leiggi.

Figure 12.c- "Allosaurus "Big Al II"" by Christian Foth[1,2], Serjoscha Evers[2,3], Ben Pabst[4], Octávio

Mateus5,6, Alexander Flisch7, Mike Patthey8, Oliver W. M. Rauhut1,2 - https://peerj.com/preprints/824/. Licensed under CC BY 4.0 via Wikimedia Commons - https://commons.wikimedia.org/wiki/File:Allosaurus_%22Big_Al_II%22.jpg#/media/File:Allosaurus_%22Big_Al_II%22.jpg

Systematic Index

Acrocanthosaurus: 29

Albertosaurus: 73, 74

Allosaurus: 5-22, 24-6, 29-31, 34-5, 48, 52, 63-4, 66-82, 99-100

Amphicoelias: 43-4

Apatosaurus: 19, 33-4, 45-6

Archaeopteryx: 61-2

Brachiosaurus: 33, 46-7

Camarasaurus: 46-7

Camptosaurus: 38-9

Carcharodontosaurus: 29

Ceratosaurus: 33, 49

Coelurus: 33, 50

Compsognathus: 93

Diplodocus: 19, 33-4, 43-5, 47

Docodon: 60

Drinker: 38

Dryosaurus: 33, 38

Elaphrosaurus: 49

Epanterias: 8, 19, 25

Fruitadens: 41-2

Fruitafossor: 61

Fukuraptor: 27

Gargoyleosaurus: 40

Giganotosaurus: 29

Gorgosaurus: 70

Guanlong: 24

Hamites: 57

Labrosaurus: 19, 80

Leedsichthys: 57

Liopleurodon: 58

Mapusaurus: 22

Megalosaurus: 51

Metriacanthosaurus: 23

Monolophosaurus: 24

Mymoorapelta: 40

Nanosaurus: 38

Neovenator: 22, 27-8

Orkoraptor: 27

Ornithomimus: 19

Othnielia: 38

Prismatoolithus (oogenus): 64

Saurophaganax: 7, 25-6

Shaochilong: 22, 29

Shidaisaurus: 22

Sinraptor: 243-4

Stegosaurus: 19, 33-4, 39-40, 70-1

Stenopterygius: 59

Stokesosaurus: 50

Supersaurus: 43

Suuwassea: 43

Tanycolagreus: 50-1

Tarbosaurus: 50

Torvosaurus: 26, 33, 52, 64

Triceratops: 19

Tyrannosaurus: 29, 50, 62-3, 75-6

Yangchuanosaurus: 24

General Index

Air Sacs: 12, 61

Ammonites: 56-7

Big Al Two: 100

Birds: 12, 37, 48-50, 61-3, 66-8, 75, 78-9

Bone Wars: 17

Countries (excluding USA);

    Argentina: 28-9

    Brazil: 22, 28

    Canada: 73, 97

        Alberta: 30, 73, 97

    China: 22, 24, 29, 55, 66

    England: 23, 51

    Germany: 59, 61-2, 93

    Japan: 27

Mongolia: 20

Poland: 20

Portugal: 6, 52, 64

Cretaceous: 22, 26-9, 32, 39, 41, 48-50, 55-6, 59

Dating;

Biostratigraphy: 30, 56

Radiometric: 30

Dinosaur Sites;

Cleveland-Lloyd Dinosaur Quarry: 7, 19, 33-4, 63

Dinosaur National Monument: 33-4

Eggs: 39, 64

Embryos: 64-5

Feathers: 50, 61, 66

Jurassic: 6, 22-4, 26, 28, 30, 34-5, 47-51, 53-5, 57-8, 96

    Kimmeridgian: 6

    Oxfordian: 6, 30

Mammals: 41, 60-1, 66, 73, 78

Marine Reptiles: 58

Mesozoic Marine Revolution: 56

North America: 6, 19, 29, 54 *(see also; Canada & US States)*

Pangea: 53-4, 56

Plants: 41-55

Popular Media: 21, 78

Pterosaurs: 59-60

Rock Units;

    Brushy Basin Member: 32

    Cloverly Formation: 32

    Fox Mesa: 63

Morrison Formation: 6, 18, 30-7, 39-41, 44, 46, 48, 50, 52, 60, 63, 91, 96

Salt Wash Member: 32

Summerville Formation: 32

Sundance Formation: 32

Sauropods: 26, 37-8, 42-4, 47, 72, 76

Sharks: 28, 57-8, 67

South America: 22, 27-8

Sundance Sea: 32, 44

Theropods: 6-7, 12-3, 20, 22-3, 26-9, 33-5, 38, 47-52, 61, 63-4, 69-70, 73, 75-8, 80-2, 91

Tyrannosaurids: 24, 26-7, 50-1, 61-3, 69, 75, 77

US States;

    Colorado: 18, 32-3

    Idaho: 30

    Nebraska: 30

New Mexico: 6, 30

Oklahoma: 8, 26

Utah: 33-4

Wyoming: 5-6, 32-3, 63, 96-8

www.ingramcontent.com/pod-product-compliance
Lightning Source LLC
Chambersburg PA
CBHW060900170526
45158CB00001B/435